A Guide to SPSS/PC+

W9-BMH-342

Neil Frude

A Guide to
SPSS/PC+

Springer-Verlag New York Inc.

© Neil Frude 1987

All rights reserved. No part of this
publication may be reproduced or transmitted,
in any form or by any means, without permission.

First published 1987 by
MACMILLAN EDUCATION LTD
London and Basingstoke

Sole distributors in the USA and its dependencies
Springer-Verlag New York Inc.
175 Fifth Avenue,
New York, NY 10010
USA.

Printed in Great Britain

ISBN 0-387-91312-2 Springer-Verlag New York

Library of Congress Cataloging-in-Publication Data
Frude, Neil.
  A guide to SPSS/PC+.
  Includes index.
  1. SPSS/PC+ (Computer system)   2. Social
sciences—Data processing. I. Title.
HA32.F78   1987   300'.285416   87-16711
ISBN 0-387-91312-2

SPSS/PC+, SPSS/PC, SPSS/PC+ Tables, SPSS/PC+ Advanced
Statistics, SPSS/Pro, SPSS, and SCSS are the trademarks of
SPSS Inc. for its proprietary computer software.

A Guide to SPSS/PC+ is not sponsored or approved by or
connected with SPSS Inc. All references in the text of this
book to SPSS/PC+, SPSS/PC, SPSS/PC+ Manual, SPSS/PC+
Tables and SPSS/PC+ Advanced Statistics are to the
trademarks of SPSS Inc.

IBM, IBM PC, IBM PC/XT and IBM PC/AT are the registered
trademarks of International Business Machines Corporation.

WordStar is a registered trademark of the MicroPro
International Corporation.

dBase II and dBase III are trademarks of Ashton-Tate Inc.

VAX is a trademark of Digital Equipment Corporation.

Contents

Preface

SPSS (The Statistical Package for the Social Sciences) is a computer program which enables data from surveys and experiments to be analyzed fully and flexibly. It has facilities for the extensive manipulation and transformation of data, and includes a wide range of procedures for both simple and highly complex statistical analysis. It also provides the opportunity for the researcher to produce fully labelled tables and graphs which can be easily incorporated into a final project report.

Over the 20 years since it was first devised, the versatile SPSS system has become an indispensable tool for many workers in social science research (including psychology, sociology, politics, human geography, business management, etc.) and in business and government. Many of the largest and most important surveys in the past two decades have been analyzed using one or other version of the system. SPSS is regularly used by government agencies, and by many major industrial corporations, market research companies and opinion poll organizations.

For many years SPSS could be run only on large (mainframe) computers of the kind found in the specialist computer installations within universities and large corporations. Advances in the speed, power and memory of microcomputers, however, have recently made it possible to produce a powerful version of SPSS for use on the desk-top machines of the **IBM PC** (personal computer) family.

When the IBM corporation introduced its microcomputers (first the 'PC' and later the PC/XT and the PC/AT) it effectively laid down new industry standards, and a number of manufacturers have since produced a whole range of IBM compatibles or IBM clones. These are able to run the same programs (software) as the **IBM PC** machines. SPSS/PC+ (and the slightly earlier version SPSS/PC) runs on machines which are equivalent to either the '**IBM PC/XT**' or the '**IBM PC/AT**'.

Familiarity with SPSS/PC+ will do more than merely enable students to analyze their current projects. Hands on experience with the SPSS system will help them to develop valuable statistical skills and will provide a working knowledge of a system which seems assured of a central role in social science research, and many fields in which surveys are employed, for the foreseeable future. When you have worked through this book you can feel confident about having a basic knowledge of a package which is the best of its kind, and the skills you will have gained are likely to remain useful in a range of contexts, and however advanced your project work becomes.

Unfortunately, there is a problem. Although SPSS/PC+ is not difficult to **use**, many beginners find it difficult to **learn to use** the package. Although the documentation for SPSS is excellent in many respects, the encyclopaedic nature of the manuals can prove daunting to those wishing to become familiar with the system. While the standard *SPSS/PC+ Manual* is an indispensable reference text, it does not attempt to *teach* the use of SPSS/PC+. Although a software tutorial sequence is included with the program, this provides a very limited introduction and is often regarded as too condensed to offer a satisfactory introduction to the system.

There are several reasons for regretting the fact that beginners often experience some difficulty in getting acquainted with the SPSS system, and that teachers often prefer to introduce the student to relevant areas of data analysis by employing less sophisticated packages:

- The power and flexibility of SPSS/PC+ means that the analysis of even modest data can be considerably enhanced through its use
- Experience with SPSS/PC+ will give all students a chance to explore the use of advanced multivariate techniques employed in many of the major studies about which they read
- SPSS has become a standard package within the social sciences and many related fields, so that students experienced in its use can feel confident that they have added to their range of research- and employment-related skills.

This book aims to teach the use of SPSS/PC+ as clearly and as simply as possible. It is a stand alone text covering many of the basic SPSS/PC+ functions and may be used by the student working independently or may form the basis for class teaching. In either case it is assumed that the student has regular access to an SPSS/PC+ facility and will be able to work through parts of the text while interacting with the microcomputer.

Although this book does not assume a prior knowledge of survey work, statistics, computer programming or report writing, it is clear that one introductory book cannot teach all of these skills. Although this book considers aspects of survey methodology (particularly questionnaire design), for information on statistics the student should consult relevant texts (including the SPSS Statistical Primer). This book is not concerned with computer programming (although it is concerned with computer **use**). SPSS/PC+ does not require the user to write programs. It is a computer program which enables the user to enter data, to describe, label and manipulate the data, and to perform simple and complex analyses. The various facilities offered by the program are activated be issuing commands which consist largely of keywords linked together following simple rules of syntax.

In writing this book I have placed the emphasis on simplicity and clarity rather than on exploring the full complexity of the system. When you have worked through this book (with hands on experience of SPSS/PC+, using either your own data or that supplied in Appendix E) you will be well placed to cope with and benefit from the information given in the *SPSS/PC+ Manual*. This book is not intended to replace that manual and cannot do this. The *SPSS/PC+ Manual* must remain the ultimate reference. For simple analyses, however, this book can act as a stand alone text.

Unlike the *SPSS/PC+ Manual*, this book provides a progressive tutorial. The simplest and most fundamental concepts are introduced first. Later sections build on the knowledge which has been acquired at earlier stages. After a brief overview of the SPSS system, the reader is taught how to use a microcomputer to explore SPSS/PC+. Experience has shown that students often find the various types of file used by SPSS/PC+ a source of some confusion. Special care has therefore been taken to distinguish carefully between the different types of file, and they are introduced progressively through the text.

This book will focus on data collected in a questionnaire survey, although SPSS/PC+ can also be used to analyze data from other types of study, including experiments. Step by step instructions will be given for conducting a survey and using SPSS/PC+ to analyze the data. The process leading up to the writing of a report can be thought of as comprising four stages. As you work through this book you will be given instructions on how to:

collect data and code it into numerical form
enter numerical data into a computer and save it
annotate the data with descriptions and labels
use the annotated data to produce summaries and tables, and to perform statistical procedures and tests

Those who are familiar with questionnaire design and data coding (including those who have experience of another version of SPSS) may prefer to begin reading at the point at which the particulars of the SPSS/PC+ system are introduced, in Chapter 3.

Chapter 1 provides a very concise overview of SPSS.
Chapter 2 shows how the data which will form the raw material for SPSS analysis (using the example of the questionnaire survey) can be collected and coded.
Chapter 3 introduces the SPSS/PC+ system as it is implemented on a

microcomputer. Details are provided of how the system can be accessed and how the user may switch between levels of program.

Chapter 4 takes the reader, step by step, through the procedures for entering data into a computer file for later use by SPSS/PC+.

Chapter 5 shows how data files may be enhanced with definitions and descriptors so that the data is fully specified to the system.

Chapter 6 discusses in detail two of the many statistical procedures which form the heart of the SPSS system. These simple procedures allow the user to examine the basic features of the set of data.

Chapter 7 introduces two further statistical procedures.

In Chapter 8 the reader is shown how data can be transformed, re-ordered and selected for analysis.

Chapter 9 examines two further types of file which can be used when conducting SPSS/PC+ analysis.

Chapter 10 provides details of another statistical procedure – used to determine correlations – and introduces a procedure which can be used to plot frequency data in graphical form.

Chapter 11 reviews the types of SPSS/PC+ file so far discussed and introduces two useful additions.

Chapter 12 returns to the topic of statistical procedures and adds analysis of variance to the range of analyses which the student can perform.

Chapter 13 provides a detailed guide showing how SPSS/PC+ can be used to perform a wide range of non-parametric statistical procedures.

Chapter 14 demonstrates the use of highly flexible and powerful procedures which can be used to present fully labelled tables to report the structure of the data.

Chapter 15 describes how two or more related files can be joined together.

Chapter 16 describes alternative ways of coding and entering data. It also discusses how portable files can be be created and transported from one computer to another. Finally, there are brief descriptions of some of the facets of the SPSS/PC+ system which are beyond the scope of this introductory book.

A number of appendices are also provided, and a guide to these is presented on pp. xiv–xv.

One way to use this book most effectively would be to devise a small study of your own – to construct a questionnaire and carry out a pilot survey (following Chapters 1 and 2), to create a **data file** and an appropriate SPSS/PC+ **data definition file** (following Chapters 4 and 5) and then to proceed to an analysis of your data (Chapters 6 to 13) before exploring some of the more complex aspects of the system (Chapters 14 to 16). When you have worked through this book you will be in a good position to devise a more sophisticated survey and to explore more complex forms of SPSS/PC+ analysis with the aid of the *SPSS/PC+ Manual*.

As an alternative to performing your own survey you might choose to make use of the data from the survey which was performed to provide many of the worked examples in this book. This survey concerns gender attitudes and was conducted on British psychology students. The questionnaire was specially constructed so that the data might allow various aspects of the SPSS/PC+ system to be explored. The original questionnaire is reproduced as an appendix (Appendix C), together with information reflecting the preparations of the data for SPSS/PC+ analysis. The **complete** set of data is also reproduced so that students, if they wish, may work through many of the examples given in the book, following the various procedures on their own computer and exploring new ways of analyzing the data provided.

A guide to the Appendices

Appendix A will be needed only if you face the task of installing SPSS/ PC+ on a microcomputer. Most readers will not have to do this – the system will already have been set up – and will not need to consult this Appendix.

Appendix B concerns the use of editor programs. These are programs which allow the user to create files for use by SPSS/PC+ and to edit files produced by SPSS/PC+. One such editor program, REVIEW, is included within the SPSS/PC+ system and its use is recommended. There may be practical advantages, however, of using another editing system (for example, the word-processing package WordStar, or a data base management system such as dBaseII), and this appendix provides a general overview of the use of such alternative editors.

Appendix C introduces the gender attitudes survey and includes the original questionnaire for this study.

Appendix D reproduces the **codebook** which was produced to provide a guide to the **variables** represented within the gender attitudes questionnaire.

Appendix E contains the full set of data from 50 (real!) respondents.

Appendix F contains the **data definition file** which can be used to make the gender attitudes data file accessible to SPSS/PC+.

These four appendices, C to F, therefore constitute a complete package. You can create your own copy of the set of data and echo the various analyses given in the text. You can also explore the data in further ways to produce your own novel analyses. Do not attempt to create the relevant computer files, however, until you have read the appropriate chapters in this book.

If you are conducting your own survey you may wish to consult various aspects of the gender attitudes survey package as a model for your own work.

Appendix G is a comprehensive glossary of the terms used in this book. It includes explanations of statistical terms, relevant computing terms and the keywords of the SPSS/PC+ system. Whenever you encounter a word with which you are unfamiliar, or are unsure about the meaning of a term, you should consult this appendix.

Appendix H provides a table which acts as a guide to features of the various types of **file** which are encountered in the use of SPSS/PC+.

Appendix I gives a brief overview of the differences between SPSS/PC+ and two other of versions of SPSS – the earlier microcomputer version SPSS/PC, and the current version used on large mainframe computers – SPSS^X.

Finally, Appendix J provides a brief synopsis of the main SPSS/PC+ commands. It will provide you with a handy reminder of the syntax of commands, and of some of the options available with particular SPSS/PC+ commands. It should be used only after you have become familiar with the use of a command by reading through the relevant material in the main text.

Acknowledgements

Many people, including students, colleagues, research workers and computer advisors have helped me in the task of creating this book. I approached some of them because of their expertise. Others were selected as 'naive subjects' and introduced to SPSS for the first time. They worked through early drafts of the book, sitting before an Olivetti SP microcomputer, and made sure that what I said would happen *did* happen. Sometimes it did not, more often because of my mistake than theirs. Students were also asked to comment on difficulties they encountered in understanding the text, and some obviously relished the opportunity for role reversal which this offered. I amended the text in response to each critical note (including 'could be worded more elegantly' and 'are you sure you mean this?').

The 'experts' found bugs of a different species, and suitable pest control measures have been taken to try to eliminate these.

I am most grateful to all those who gave their time so generously. The principal helpers are listed below in alphabetical order. I will leave it to these individuals to judge whether I approached them for expert or 'grass roots' advice: Annette Davies, Hilary Griggs, Rosemary Jenkins, Pamela Kenealy, Ray Kingdon, Mark McDermott and Martin Read.

Function Keys and Their Use

The function keys on the computer keyboard are labelled <F1> to <F10> or <F18>. Several of these have special uses within the SPSS REVIEW procedure. They often have to be pressed in conjunction with the <SHIFT> key, indicated by ↑, the <CONTROL> key, indicated by ^, or the <ALT> key, indicated by a. Thus the function key <F> has different functions when pressed alone or with any of these keys. The instructions will be written as <F4>, <↑F4>, <^F4> or <aF4> respectively.

The <RETURN> key on your keyboard may be indicated by a bent arrow symbol.

1 Introducing SPSS/PC+

This chapter will lead you gently through the four stages involved in conducting an SPSS analysis. It provides an overview, and few details will be given at this stage. It is *not* important that you try to learn the particulars of the many aspects covered, but it *is* important that you read with understanding and gain a general impression of the SPSS approach.

The four stages which will be briefly covered involve:

- collecting data and coding it into numerical form
- entering numerical data into the computer and saving it
- annotating data with descriptions and labels
- using the annotated data to produce summaries and tables, and to perform statistical tests

1.1 Collecting and coding data

To examine facts relevant to social and psychological issues, social scientists need information, or data. Some data can be gathered from published sources, but a social scientist often has to collect new data by conducting an original study. This must be specially devised to provide the desired information or to answer specific questions.

The basic atom with which the social scientist deals is the *variable*. This can be any type of information item, including sex (i.e. male or female), age, income, marital status, an attitude, a belief, a personality score, the number of children a person has, the party a person voted for at the last election, etc.

One common type of study used in the social sciences is the experiment. Here one variable (e.g. the length of a list of words to be learned) is deliberately manipulated and the effect of this on another variable (e.g. the time needed to learn the list) is measured.

Another type of study is the survey. Often this will involve inviting people to participate by giving some information about themselves or other people. The subjects (or respondents) may be interviewed or given questionnaires to complete. Certain characteristics of the respondents are noted (e.g. sex, age) and comprise some of the variables. The answers given to the questions are the other variables.

Some questionnaires may be standard tests (like personality tests) whereas others will be a series of questions put together by the researcher especially for the study.

Some interviews are relatively unstructured – more like a normal conversation – but others are structured, i.e. they closely follow a script (and are thus the verbal equivalent of a written questionnaire). The main advantage of unstructured interviews is that they yield more individual profiles of the respondents and allow interesting points to be probed in depth. The main advantage of structured interviews (or questionnaires) is that they allow the researcher to compare all the respondents' answers to the same questions given in the same order.

Within interviews or questionnaires, some questions permit respondents to answer in a free, or open, way. Thus:

'What are your feelings about the way in which this country is currrently governed?' ....

The main advantage of open-ended questions like this is that they allow a range of individual expression and thus provide rich and varied data. Unfortunately, however, open-ended questions are not easy to analyze. It is difficult to compare the answers which different respondents give to the same question.

Other questions are supplied with a fixed range of permitted responses. Here are three examples of fixed-range questions:

(i) **Do you believe in the existence of God?**   **YES   NO**
(ii) **If you had to choose a painting for your room would you choose:**
    (a) a fine print of an Old Master
    (b) an original nineteenth-century water-colour
    (c) a modern painting?
(iii) **Indicate, by circling ONE number, where you stand politically on the left–right dimension:**

	Extreme left					Extreme right
1	2	3	4	5	6	7

The main advantage of closed-ended questions (that is, where the answers have a fixed range) is that all respondents answer within the same restricted range, thus making comparisons between respondents very easy.

All data to be used by SPSS must be coded for all subjects in the same way. For each respondent the value for any variable (for example, the answer to a question) must be determined. Thus the variable *sex* has two possible valid values (corresponding to male and female). A coding scheme must be devised so that one code (say 1) corresponds to 'male', and another code (say 2) corresponds to 'female'. A variable such as *age in years* hardly needs a coding system because the form in which the data is collected (23, 56, etc.) can stand. Interviews (or written questionnaires) with closed-ended questions are easy to code because each of the range of permitted answers is assigned a particular code value (perhaps 1 for 'yes', 2 for 'don't know' and 3 for 'no'). Closed-ended questions thus yield information in a form which is ideal for SPSS analysis. Open-ended questions are likely to raise problems, however. Although it may be possible to devise a workable method of coding the data obtained from such questions, this can prove difficult in practice.

Applying the coding system devised by the researcher, all of the raw data from a survey (including the subject's identification number, sex, social class, age and the responses to particular questionnaire items) are transformed into numbers (letters can also be used, but we will not discuss such alphanumeric variables until Chapter 16). The data for each respondent is thus represented as a series of numbers, and the data from all of the respondents together forms the dataset on which SPSS/PC+ will operate.

The **dataset** therefore consists of data from a number of respondents, or cases, and the data for each case is represented by a series of numbers. These numbers represent the coded *values* (for the particular respondent) of data items relating to a number of different *variables*. Transforming raw data into values is achieved by applying the coding system.

1.2 Entering and saving data

Once all the data obtained from a study has been collected and coded into numerical form it is entered into the computer as a set of numbers. This consists of a table or matrix of numbers, like the one below (but usually a lot bigger):

```
019865976597659
029867596576586
035475486547654
047865967654445
```

Each row (there are four rows) contains the data from a single respondent. The numbers in the columns along the rows (there are 15 columns in the sample) represent the values which have been determined for each subject for particular variables (obtained from, say, the answers to a questionnaire). The whole matrix represents the dataset.

To prepare the data for entry into the computer we use a program referred to as an **editor.** We type the data across the screen as a long series of numbers. Each time we have finished typing the data for one respondent we start a new line (row) of data by pressing <RETURN>.

When the complete dataset has been keyed in we follow instructions to save the dataset and the machine then stores the data matrix as a **data file** with a particular name. It is permanently stored on a magnetic **disk.**

We can call up the data at any time to use it, to modify it or to add to it. Once our dataset has been established we never need to type it in again. Thus we can perform many separate analyses, on many different occasions, after having typed in our data matrix only once. This facility for the 'once and for all' entry of our data is one of the great advantages of working with a package rather than with separate programs for analysis.

1.3 Annotating data

So far the data has been collected, coded and entered into the machine as a data matrix. But no instructions have been given to the machine about how to interpret or label the numbers. The next stage is to describe the data so that the system can identify the variables represented within the dataset.

The first thing we do is to tell the machine how to slice the long string of numbers for any one subject. **We** know (but the system does not) that the first row of numbers:

019865976597659

represents not *one* 15-digit number but rather a **series** of smaller numbers.

The 15-digit string **might** represent 15 single-digit numbers:

0 / 1 / 9 / 8 / 6 / 5 / 9 / 7 / 6 / 5 / 9 / 7 / 6 / 5 / 9

or it **might** represent a 'mixed bag' of single-, double- and triple-digit numbers:

01 / 9 / 8 / 65 / 9 / 76 / 5 / 9 / 7 / 659

i.e.

01,   9,   8,   65,   9,   76,   5,   9,   7,   659

In this case the 15 **digits** would represent coded values for only 10 **variables.**

The instructions given to the computer on how to slice up the dataset are called format instructions. The most usual way to arrange a dataset is in fixed format, i.e. where the data for all respondents has precisely the same format. Thus the data relating to a particular variable occupy precisely the same position in the number series for all respondents and each variable is represented by a one-, two- or three-digit (etc.) vertical slice through the dataset.

Thus for all the data in the matrix displayed above, the arrangement in terms of variables (call them *V1* to *V10*) would be:

01	9	8	65	9	76	5	9	7	659
02	9	8	67	5	96	5	7	6	586
03	5	4	75	4	86	5	4	7	654
04	7	8	65	9	67	6	5	4	445
V1	V2	V3	V4	V5	V6	V7	V8	V9	V10

Obviously, the computer will need to be informed of the formatting arrangement for the dataset in order to locate variables within it. Since the same format is used for the data relating to all cases within the dataset, a single format specification will be sufficient to cover all cases. Having been provided with the format specification, the computer will be able to recognize the numerical organization within the long string of digits which comprises the coded data for each respondent.

(In contrast to this fixed format system, SPSS/PC+ permits an alternative free format system. This is used less frequently, however, and need not be discussed at this stage. An explanation will be found in Chapter 16.)

As we specify the format, we also invent names for each variable, naming them – together with their position – in the order that they occur within the dataset. For some variables there will be an obviously appropriate name – thus for the sex of the respondent we are likely to choose *sex* as the variable name. Other obvious names include *age*, *class* and *id* (for identification number).

In other cases there will not be an obvious choice of name. Suppose, for example, that a variable is the answer to the question: 'Do you believe in the existence of God?' We might name the variable *belief*, *beliefg*, *god*, *bgod* or *bg*. Names can be no more than eight characters in length, and spaces are not permitted, so *belief in god* would not be permissible as a variable name. The name chosen should be the one which will prove easiest to remember and recognize.

Sometimes it may not really be possible to capture the nature of the variable in a simple descriptive sequence of letters. In such cases we can resort to using such names as *Q14* (meaning the coded answer to Question 14) or *var025* (meaning, simply, variable 25).

Once they have been established, variable names and format specifications will rarely need to be changed. Whatever analyses and manipulations will be performed with the dataset, the same variables will occupy the same locations within that dataset. Thus a statement which names the variables and specifies their format can become a permanent descriptor of the set of data.

So far, the system has been instructed about how the series of digits in the dataset are to be broken up into slices, each representing a variable, and it has been provided with names for these variables (in the order that they occur). SPSS/PC+ also provides excellent facilities for **labelling** the variables. Labels are used to annotate the output of analyses – the screen displays and printed output produced as a result of the user's requests for analyses of the data set. If the system displayed only the names of the variables analyzed (*var025*, *var034*, *bgod*, etc.) we would probably have to look back at our notes to identify particular variables. If the output is annotated with fuller labels, however, it is easier to make sense of a table or statistical result at a glance.

Thus the system can be informed that *var025* should be labelled 'number of church attendances in past year' or that *bgod* should be labelled 'belief in the existence of God'.

SPSS/PC+ also enables us to label the values associated with any variable. Remember that data corresponding to each variable will have been coded and that each variable will have a range of valid values. Thus the raw data for the variable *soclass* (which could have the variable label 'social class') might have been coded into five categories. Each respondent for whom we have the raw data on social class will have been given a coded value (perhaps using the range 1 to 5) for the *soclass* variable. Particular values for this variable (1, 2, etc.) can now be labelled (e.g. 1 might be given the label 'manual working class'). The two valid values for the variable *sex* could also be labelled (for example, the value 1 might be labelled 'male' and the value 2 might be labelled 'female'). Other commonly used value labels include 'agree', 'disagree', 'yes', 'no', 'always', 'often', 'sometimes' and 'never'. Value labels, like variable labels, are not used to communicate with or to instruct the system. They are included merely to enable the system to annotate the output as it is displayed on the screen or printed on paper. They are not necessary for the execution of any analysis but are optional extras which add to the presentation and ease of interpretation of output.

Everything that has been described in this section can be considered as permanently describing the dataset. Information about the format (or 'slicing') specifications, the variable names, the variable labels and the value labels can be stored permanently along with the data and will be used by the computer whenever we use SPSS/PC+ to analyze the particular dataset. Such information, therefore, need be entered only once and will remain in effect unless changes are deliberately made.

1.4 Using the annotated data

We are now at the most interesting and productive stage. It will soon be time for SPSS/PC+ to work for us. Some of the tasks it performs are relatively simple (but would be extremely tedious to perform by hand on a large dataset) while others are extremely complex.

Instructions to the system are given by various commands. Some of these (**operation commands**) affect the way in which the system works. Some other commands (**data definition** and **transformation commands**) label existing variables or create new variables, and some (**procedure commands**) request that analyses be performed on the data. Although commands are generally very easy to use, there are certain rules of syntax which need to be adhered to. These will be introduced progressively through this book.

One example of a transformation command is **COMPUTE**. This is used to create (by computing) new variables from those which are already present in the dataset. Suppose, for example, that our dataset consists of students' scores on a number of class tests and their results in a final examination. We will name the variables *test1*, *test2*, *test3* and *exam*. Suppose that a student's final grade is obtained in the following way. The three test scores (in percentage terms) are equally weighted and together constitute 50 per cent of the final grade, with the examination constituting the other 50 per cent.

SPSS/PC+ can be asked to compute a new variable, *grade*, using the existing values of the four variables for each student:

```
COMPUTE GRADE= (TEST1 + TEST2 + TEST3 + (EXAM*3))/6.
```

This will yield, for each student, a value for a new variable *grade* (notice in the above example that ∗ means multiply and / means divide).

The command instructs the computer to add each student's three test scores, then to add his or her exam score multiplied by 3, and then to divide the result of this computation by 6. The new variable *grade* will thus have a value for each student (equivalent to the percentage score for the final grade).

A **procedure command**, **FREQUENCIES**, can be used to instruct the system to return the frequency of cases (in the above example each student represents a single case) which have each of the values of a particular variable. The **FREQUENCIES** command can also be used to provide a number of other basic facts about the dataset. Thus it can be used to find the average of any variable (including both original variables in the dataset – *test1*, *test2*, etc. – and new computed variables – e.g. *grade*) or the range of any variable (i.e. the difference between the lowest and the highest values for the population of cases in the sample as a whole). **FREQUENCIES** will also produce a graphical display of frequencies.

Suppose we want to obtain basic information about the variable *age* (e.g. how many 23-year olds there are in the sample, how many 35-year olds, the average age of the sample, etc.). The instruction **FREQUENCIES AGE.**

would be sufficient to obtain the average (**mean**) age of our sample, an indication of the general spread (**standard deviation**) of the ages, and the minimum and maximum age (i.e. the ages of the youngest and the oldest) of our sample. The output will provide the frequency of each age and, if requested, supply a histogram or a barchart to illustrate this frequency distribution.

(Statistical terms are explained in the glossary (Appendix G). Although no statistical knowledge is required to understand the directions given in this book for using SPSS/PC+, any proper use of the statistical procedures used in the package will of course require such knowledge. The reader is directed to one of the many statistics textbooks available, and to the *SPSS/PC+ Manual* and *Advanced Statistics Manual*.)

After obtaining an overview of the average and spread of each single variable the researcher will generally wish to examine the relationship between pairs (or groups) of variables.

Thus it may be required to examine the relationship between the variables *sex*, ('male' and 'female') and *class* (say, 'working class' and 'middle class') to see whether, within the sample of respondents, the two sexes are unequally distributed between the social classes. An unequal distribution would mean that, for our sample, there was an association between *sex* and *class*. If we were to examine this question without the use of a computer package such as SPSS/PC+ we would need to count the number of:

working class males   working class females
middle class males   middle class females

It would then be useful to arrange for a display in the form of a 2×2 table with four cells:

```
                                      SEX
                            Male        Female
                          ..............................
                          .            .            .    Total W.C.
      C   Working Class   .    45      .    23      .        68
      L                   .            .            .
      A                   ..............................
      S                   .            .            .    Total M.C.
      S   Middle Class    .    19      .    38      .        57
                          .            .            .
                          ..............................

                          Tot. Male    Tot. Female      TOTAL
                             64            61             125
```

Such a table is referred to as a crosstabulation or contingency table. The above example is a 2×2 crosstabulation (which therefore has four cells). A similar exercise could be performed if *class* were coded into three values (say, 'middle class', 'non-manual working class' and 'manual working class'). The resulting crosstabulation would be a 3×2 table containing six cells.

A variety of statistical tests can be performed with such a two-variable classification to establish whether there is a significant statistical relationship between the two variables concerned (here *sex* and *class*).

SPSS/PC+ will crosstabulate such information, and analyze it statistically, using the command **CROSSTABS**.

For the above example the instruction would be:

```
CROSSTABS  SEX BY CLASS.
```

This would produce a crosstabulation table, labelled with any variable labels and value labels which had previously been provided. The number of cases with relevant values for both of the variables would be displayed in each cell. Thus the final count of all cases in the sample which had been coded as 'male' on the *sex* variable and as 'working class' for the *class* variable would be displayed in the 'male working class' cell.

Associated with many SPSS/PC+ commands are a number of **subcommands**. These have various uses but often have the function of allowing the user to tailor the form which an analysis will take, or the form in which output will be displayed.

One of the most frequently encountered subcommands is **OPTIONS**. By quoting one or more numbers on an **OPTIONS** subcommand the user is able to control many aspects of the procedure. The meaning of these numbers changes depending on which command is being issued. Thus when used with the **CROSSTABS** command, the **OPTION 2** is specified if the user does *not* wish the variable- or value-labels to be displayed or printed. The following command (including the appropriate **OPTIONS** subcommand) would therefore be used to instruct **CROSSTABS** to display and print a table *excluding* these labels:

```
CROSSTABS  SEX BY CLASS
/OPTIONS= 2.
```

(Notice that the subcommand is preceded by a slash / . This is an example of a rule of SPSS/PC+ syntax which will be fully explained later – remember that the function of this chapter is merely to provide an overview.)

Another frequently encountered subcommand is **STATISTICS**. Although many of the SPSS/PC+ commands automatically produce an output which includes statistical information, in many cases there is also a range of additional **STATISTICS** which may be specifically requested.

Thus certain statistical tests can be requested if a **STATISTICS** subcommand is issued as part of a **CROSSTABS** command. For example, if **STATISTIC 1** is specified, then a chi-square test will be performed on the data which appears in the crosstabulation. If **STATISTIC 11** is specified than a correlation will be performed on the data. Thus if we wish to use **CROSSTABS** to print a table which excludes variable- and value-labels, and to calculate both chi-square and a correlation coefficient, we would specify:

```
CROSSTABS  SEX BY CLASS
/OPTIONS= 2
/STATISTICS= 1, 11.
```

Sometimes we would wish to split our population of respondents into two groups (for example, males and females) and test to see whether there is a difference between the groups on a variable such as *age*. This, too, can be easily done. The appropriate command for such a task would be **T-TEST**. To find out whether there is a significant age-difference between the males and the females in our sample we would need to use the variables *age* and *sex*.

We would need to specify the two groups – males (coded 1 on *sex*) and females (coded 2), e.g.:

```
GROUPS= SEX(1,2)
```

We would then specify any variable or variables on which we wished to compare our groups, using:

```
VARIABLES= AGE
```

Since a *t*-test analysis is required we would use the **T-TEST** command, and putting the various elements together using the appropriate SPSS/PC+ syntax we would enter the command:

```
T-TEST   GROUPS=SEX(1,2) / VARIABLES= AGE.
```

If we wished to perform *t*-tests to compare males and females on age, length of education (say, *leduc*) and yearly salary (say, *income*), this could be specified in a single command:

```
T-TEST   GROUPS=SEX(1,2)/VARIABLES= AGE LEDUC INCOME.
```

For those who are statistically sophisticated, let me add, to whet your appetite, that SPSS/PC+ provides for a whole range of 'non-parametric' analyses, analysis of variance (anova), correlations, and regression analyses, etc. It will also plot data graphically in a variety of ways.

For the **really** sophisticated I will mention, too, that the optional *SPSS/PC+ Advanced Statistics* package will also perform – with equivalent ease of operation – factor analyses, discriminant and cluster analyses, and log-linear modelling.

And if such terms mean little to you at this time, relax! SPSS/PC+ will allow you to explore your data to the full, and with instant feedback you will be able to fully explore your dataset and become familiar with its properties as few were able to do before the advent of microcomputer based versions of the SPSS system. Your statistical literacy and your statistical skills will increase, and the depth and quality of your analyses and report presentation will improve significantly.

Remember that the purpose of this initial overview is to show how simple it can be to use various procedures. Do not try to learn the **details** at this stage – and do not worry if the statistical terms (like *t*-test) mean little to you. The aim at this stage is to give a general idea. Those who are more sophisticated in statistics, however, may develop a special appreciation of how easy it is, once a dataset has been established, to perform otherwise complex analyses.

By now you should appreciate the general concepts underlying SPSS/ PC+. It is **not** important that you **remember** the details contained in this chapter, but it **is** important that you have **understood** all of the material. In the following chapters we will be building on this basic understanding.

You should now have a preliminary understanding of how:

- data is collected and coded into numerical form
- numerical data is entered into the computer and saved
- the data is annotated with descriptions and labels
- the annotated data is used to produce summaries and tables, and to perform statistical tests

If you feel unsure about any of the general concepts it would be well to re-read one or more of the sections before going on to the next chapter. Remember, too, that you might find help on any particular difficulty by consulting the comprehensive glossary in Appendix G.

When you are confident that you have a general grasp of the SPSS/PC+ system, as outlined in this chapter, it is time to proceed to Chapter 2.

2   Collecting and Coding Data

This chapter deals with the first stage of the four-stage paradigm outlined in Chapter 1. In this chapter we learn how to collect data and code it into numerical form. The discussion will include a consideration of:

- variables, cases and values
- coding strategies
- preparation of the questionnaire and codebook
- coding the dataset

2.1 Variables, cases and values

Typically, SPSS/PC+ will be used to analyze data from a survey in which we have collected many items of information (**variables**) from (or about) a number of different people (called **subjects** or respondents or, more generally, cases). Alternatively, our data might consist of variables associated not with individual people but with other kinds of cases – different products, for example, or different countries. The aim, in collecting and coding data, is to arrive at a situation in which we have, for each **case**, a **value** for that case on each **variable** in which we are interested.

Thus for a **variable** which has been given the name *age* we might have the value **35** for case 1 (John Doe) and the value **23** for case 2 (Sarah Smith). For the variable *income* we might have the value **12000** for case 1 (John Doe) and the value **09500** for case 2 (Sarah Smith).

The data to be analyzed is organized into a grid with each case as a separate **row**, and each variable as a **column** of the appropriate width (see Table 2.1).

Here each **case** is represented by a row of data. Each **variable** is represented by a column which is either two single columns or four single columns in width. For each case, a **value** is associated with each variable, and these values (16 in the example provided) constitute the **data** for analysis.

Table 2.1

Case	Variable 1 Age	Variable 2 Income	Variable 3 Height	Variable 4 Weight
001	35	1750	72	61
002	23	1200	68	41
003	25	2100	70	54
004	27	2335	63	38

A typical SPSS/PC+ project may involve hundreds of cases, and up to 200 variables may be represented within a single dataset.

2.2 Coding strategies

In most of the examples provided in this book, the data analyzed comes from a survey in which a number of students (cases) provided information in response to a number of questions. The gender attitude survey which provided the data for many of the examples in this book is briefly described in Appendix C. Each questionnaire item is treated as a separate variable and each subject's response to the item has been coded into an appropriate value.

Although SPSS/PC+ can provide some limited analysis of alphabetic data, it will be assumed here that the answers to all of the questions (i.e. all of the information to be analyzed) can be coded into numerical form. Some data will actually be **collected** in numerical form (e.g. age in years, the number of children in the family, or the degree of agreement expressed towards a statement on a defined scale of, say, 1 to 7 points). Information on the use of alphanumeric variables and variables which are non-integer (i.e. which contain decimal points) is provided in Chapter 16. Other data will not be collected in numerical form but will have to be coded later into a numerical system. Thus sex (**male** or **female**) may be coded numerically so that one number (say, 1) signifies 'male' and another (say, 2) signifies 'female'. Such numerical coding can be achieved for almost all data. Thus for a question to which respondents are invited to answer 'Yes', 'Not sure' and 'No', we could code the answers as 1, 2 and 3 respectively. The actual coding system used to translate answers into numerical values may be arbitrary, but we must keep a clear record of the key used to translate the responses into numbers. For coding any particular variable (such as the answers to one particular questionnaire item) we must apply precisely the same coding system to each case (or respondent).

Thus, in coding the sex of a respondent into a numerical value, instead of

1 = 'male', 2 = 'female' we could equally well use 1 = 'female' and 2 = 'male'. But, once the rule has been established, the sex of **all** subjects must be coded according to the **same** rule.

Another example of numerical coding is provided by the case where we ask respondents how much they agree or disagree with a statement using a fixed range of shades of agreement. Thus:

'Do you agree that women should have the same rights as men?'

Respondents could be asked to indicate how much they agree or disagree with this statement by ticking one of five responses:

> **strongly agree**
> **agree**
> **neither agree nor disagree**
> **disagree**
> **strongly disagree**

For analysis, the answer each respondent gives to this question could be coded numerically (for example, using 1 to indicate 'strongly agree', 2 to indicate 'agree', 3 to indicate 'neither agree nor disagree', 4 to indicate 'disagree' and 5 to indicate 'strongly disagree').

So far, each of the examples has involved a 'forced choice' question. Another example would be 'Would you prefer (a) a restful holiday, (b) an active holiday or (c) something in between?' We could code the responses (a), (b) and (c) as 1, 2 and 3.

Sometimes, however, we might prefer to allow an open response. Thus:

'What is your favourite hobby?' (please state) ....

Here we are inviting (and will get) a huge variety of answers, from alchemy to zoo-visits. We have to code these numerically, and can do this in a variety of ways. Take two extreme examples:

> (a) We could simply decide to code any hobby as either 'outdoor' (code 1) or 'indoor' (code 2).

or

> (b) We might decide to wait until all questionnaires are returned and then list all the answers obtained. We could then provide a coding number for each hobby mentioned as the favourite hobby by at least one person. Thus if 57 different favourite hobbies were produced they could each be given a code number (01 to 57).

Between these two extremes we might decide (either before the survey or when we have had a chance to look at the range of answers) a number of categories for coding. Thus we might divide the favourite hobbies into the following categories:

1. Sports/games   4. Crafts
2. Social   5. Music
3. Collecting   6. Other

We could then code the data using the numbers 1 to 6. Sometimes a difficult choice would have to be made. For example, should 'collecting and listening to Beatles albums' be coded as 3 or 5? Because only one value (or code) is permitted for each respondent on each variable it would *not* be possible to use both. Notice also the 'other' category, used for the eccentric cases which don't fall easily into one of the set categories. Some such miscellaneous category is usually needed when we attempt to code responses to open-ended questions.

There is another issue which needs to be raised. It is inevitable in a questionnaire survey that some respondents will fail to return a complete set of responses. A subject may make the error of missing out an item or may turn two pages at once. Subjects sometimes object to a question and refuse to answer, or write an uncodeable (and sometimes unprintable) comment rather than choosing from a range of responses provided. In all such instances the data is treated as missing. The way to deal with what would otherwise be a hole in the dataset is to provide a special extra value for coding missing data. Any missing data is then coded with this missing value. For example, if *sex* is coded as 1 for 'male' and 2 for 'female', the value 0 (or perhaps 3 or 9) can be assigned as the missing value.

As part of the process of defining the dataset, the system can then be instructed that a particular value for a variable represents not real data but missing data. The SPSS/PC+ system recognizes such an instruction and can deal appropriately with missing data. (For example, if it has been instructed to treat 99 as the missing value for the variable *age* it will *not*, in calculating the average age of the respondents, include any 99 values as if there were people in the sample who were just one year short of a century!)

Finally, note that some open-ended questions will be so varied that any simple coding organization will be difficult to devise. For example,

'What do you remember of your first day at school?'.................

The data which will emerge from such an open-ended question will often be rich and interesting, but they will not be in an ideal form for statistical analysis.

Contrast the above open-ended question with the following:

'How do you recall your first day at school?' (tick ONE)

.... **interesting**

.... **boring**

.... **neither boring nor interesting**

.... **cannot remember**

This is easily coded, and therefore will be easily amenable to statistical analysis (although it has to be admitted that the data which will emerge will not be nearly as full or interesting, in this case, as that which would be likely to emerge from the open-ended question).

The raw data from written questionnaires, interviews, etc. is therefore prepared for use by SPSS/PC+ by being coded according to a uniform coding scheme. All of the information is thus reduced to a series of numbers.

Some of these numbers will be real numbers (like age in years) and represent numerical variables. Other numbers, however, will be code numbers (like 1 = 'yes', 2 = 'no', 9 = 'missing'). These represent the data from categorical variables, and the codes given to the various categories are determined by a key devised specifically to code the information relating to the particular variable.

The user provides the computer with the dataset in the form of a matrix of numbers (where the columns are organized into single-column, two-column or larger vertical slices, each representing a single variable, and where each row – or for extensive datasets each group of two or three rows – corresponds to a particular case). The variables may be derived from questionnaire items of varying style and format, but the information relating to each variable must be coded into numerical form and placed in the appropriate column(s) of the dataset. Thus, for a sample of cases, each variable corresponds to a vertical slice through the dataset (or data matrix).

2.3 Preparation of the questionnaire and codebook

It is useful to develop a questionnaire in conjunction with a **codebook** – your own guide to the variables and the way in which they should be coded. Some of the codebook information can usefully be added to the final version of the questionnaire itself (in a separate column with a message to the respondent, such as **PLEASE LEAVE BLANK**). The responses to the completed questionnaire can then be coded directly on the form which has been completed by the respondent, and later copied onto special **data sheets**. From the data sheets the data can be entered directly into a computer **data file**.

Preparing a questionnaire

The first step is to construct an initial draft of the questionnaire. It is useful, even at this stage, to think in terms of the variables which will be produced. Certain types of analysis demand **numerical** rather than **categorical** variables, for example, so the form a question takes may be partly determined by the type of analysis needed. The questionnaire should be comprehensive enough to cover the full range of variables required, without being over-long. It should have a definite logical structure, and the order of questions should be carefully considered. Branching may be included (e.g. **'If YES then go straight to Question 7 – otherwise answer Questions 5 and 6'**), but such complexities should be kept to a minimum in questionnaires which are to be completed in writing by the respondents (and should be avoided altogether, if at all possible).

The user needs to decide whether questions are to be open-ended or closed, and any closed questions demand careful consideration regarding the range of alternatives to be provided. If only one answer is to be permitted to a question, the alternatives presented should be **mutually exclusive** (i.e. only one of them **can** apply in any one case). The alternatives should also be **mutually exhaustive** (i.e. one or other of them **must** apply to each case).

An example of a question which breaks both of these rules is:

'What is your current marital status?'
    married
    divorced
    previously married
    widowed

This question does not provide mutually exclusive alternatives because a divorced person could validly tick both 'divorced' and 'previously married'. For a general population, the alternative answers would not be mutually exhaustive, either, because no provision is made for a response from a single (= 'never married') person.

Another golden rule in questionnaire design is to avoid ambiguous phraseology. The question **'Were you married last year?'** for example, could mean **'Did you *get* married last year?'** or **'Were you of married status last year?'**. Thus a person who had been married for the past 10 years would answer NO in response to the first meaning or YES in response to the second meaning.

Some of the more subtle ambiguities (including words or phrases which have different shades of meaning for different people) can only be identified as the result of piloting an early draft of the questionnaire with respondents similar in age, social class, and educational level to those who will form the real survey population. The research worker needs to talk at some length with people who have attempted to complete the questionnaire, to discover any problems they may have had in understanding or answering the questions.

Before the final version of the questionnaire is typed or printed, and multiple copies made, it is important to consider whether provision is to be made for coding the data on the questionnaire itself. If on-form coding is to be performed, the inclusion of information from the codebook might aid the task of the coder. Such information, and any spaces provided for coding, should be clearly separated from the part of the form which the respondent is expected to complete. A line should be used to separate the two fields (the respondent field and the coder field) and the coder field should be headed with a '**PLEASE LEAVE BLANK**' message.

The gender attitudes questionnaire (Appendix C) provides an example of how a questionnaire may be prepared to facilitate easy on-form coding. Such aids to coding, however, can be included only after the codebook has been compiled.

Preparing a codebook

When the questionnaire itself has been prepared in its final version (apart from any coding provisions which are to be included), it needs to be examined carefully and a codebook prepared. The codebook provides a guide to the variables to be extracted from the questionnaire. It will usually contain the variable **names**, the valid values of data for each variable (and a value to be assigned to data which is missing), and the nature of the data (e.g. **C** for categorical variables – such as *class* or *sex* – and **N** for true numeric variables – such as *age in years* or *income*). In addition, variable **labels** (remember that these are not the same as variable **names**) and value labels (e.g. 1 = 'male', 2 = 'female') may be included for at least some of the variables.

The questionnaire first needs to be **itemized**. Each item will yield one **variable**. In most cases there is no problem. Many questions will produce a single variable. Thus:

   (a) **'Women should have the same rights as men'**
        strongly agree
        agree
        neither
        disagree
        strongly disagree

   (b) **'Current occupation'**...............................
        (Later coded in terms of social class)

(c) 'Which of the following is your favourite colour?'
red
orange
yellow
green
blue
indigo
violet

(d) 'For how many hours in an average week do you watch television?'..........................

(e) 'Which ONE of the following games would you prefer to engage in?'

| Hockey ... | Football ... | Skiing ... | Squash ... |
| Tennis ... | Shooting ... | Golf ... | Chess ... |

In each of these examples only *one* answer is permitted – and thus each answer can be coded as a single variable. Consider, however, a variation of example (e):

(f) 'Which of the following games have you engaged in during the past MONTH?'

| Hockey ... | Football ... | Skiing ... | Squash ... |
| Tennis ... | Shooting ... | Golf ... | Chess ... |

Here some respondents may tick **none** of the games listed. Others may tick **one**, and some will tick **more than one**. We could code the answers to this question in terms of a single variable by merely counting the number of ticks given – 0 to 8 – but this form of coding would lose some of the precision or fullness of the data collected. We might wish, at some future stage, to count the number of our respondents who had engaged in *each* of the individual games. To preserve the full information from respondents' answers we should treat this question as containing eight separate items, and hence code the answers in terms of eight separate variables.

Thus the variable *hockey* might be coded 0 if there was no tick present, and 1 if a tick had been given (and so on for the other seven games). If we later wished to count the total number of listed games engaged in by a particular respondent during the month, we could use a special **COMPUTE** feature of the SPSS/PC+ system to add the number of ticks given to the list (i.e. the number of variables, among the eight, which were coded with a 1, meaning 'game engaged in'). If we had opted to include all of the information from the question in terms of a single variable, however, and merely counted the ticks, there is no way in which we could later unscramble the resulting number (0 to 8) to determine how many respondents had engaged in, say, hockey. It is worth bearing in mind the old saying 'You can't unscramble eggs'. The best policy is to include the data in as full a form as you will possibly require, and this might mean that a single question will yield not one but several variables.

After itemizing the questionnaire, and thus producing a list of variables with suitable names (each up to eight characters in length), we should then consider, for each variable in turn:

(i)   the number of columns which the variable will occupy within the data matrix
(ii) the position of these columns
(iii) the variable type
(iv) the range of valid values for the variable
(v)   the value labels to be assigned (if any)
(vi) the value to be assigned to any data which are missing
(vii) the variable label to be assigned (if any)

(i) **Data columns**: With a **C** – categorical – variable, if there are fewer than 10 valid codings then a single column will accommodate all possible values for that variable (for example, codings 1–9 may be used for 'valid data', with 0 reserved for 'missing' data). If there are more than nine valid values then two columns will be needed to accommodate the variable; if there are more than 99 valid values three columns will be needed, etc.

With **N** – numerical – variables (like *age*), the number of columns occupied will depend on the range of values possible. Thus, unless working with an unusually elderly population, two columns should accommodate *age* (in years). *Income* (annual, in pounds sterling or dollars) may require five or six columns. *Height* (in inches) will require just two columns; in centimetres, three columns. If more precise data is to be used, including numbers beyond the decimal point, then more columns will need to be specified. (For information on decimal data, see Chapter 16.)

(ii) **Column position**: Once we know the number of columns needed to accommodate the variables in our list we can begin to assign particular columns within the data matrix to particular variables. Thus if the first variable, *id* (identification number), is assigned three columns, these can be columns **1** to **3** (inclusive). The next variable (say, *sex*, accommodated in just one column) can then be placed in column **4**. We can therefore map out the full specification of the data matrix by following the sequence of variables precisely, and assigning at each stage the next column for a single-column variable, the next two columns for a two-column variable, and so on. The column(s) occupied by the variable should be entered in the codebook.

SPSS/PC+ works with an 80-column format which has become the standard for much computer work. If, when assigning columns, you approach column **80** you should stop at a suitable column number (say **77**, **79** or **80**), taking care that data for a single variable will **not** begin on one line and end on the next. Since the first data line is full (or as full of complete variables as it can be) you should now begin assigning columns on the second line. Start numbering again at column **1** and make a note (within the codebook) that the column numbers now represent a second data line *for the same case.* Sometimes three, four or even more lines may be needed to accommodate the data for a single case.

(iii) **Variable type**: Different kinds of analysis are appropriate to cat-egorical data (such as that relating to country of origin, sex, or favourite sport) and numerical data (such as that relating to age in years, annual income or number of children). It may prove useful to be reminded of the type of data included for each variable. *Age*, for example, could be represented numerically ('age in years', 01 to 99) or categorically (1 = 'young' = 30 or younger; 2 = 'middle' = 31 to 50; and 3 = 'old' = 51 or over).

Placing a **C** or **N** in the codebook, beside each variable, may act as a useful reminder of the variable type at a later stage when complex analyses are being planned.

It should be noted that it is not always easy to decide whether a variable should be considered as categorical or numerical. For example, some variables may include the responses to five-point or seven-point scales, where the lower end signifies, say, complete agreement and the upper end complete disagreement with a statement. Some people would argue that the fact that numbers, rather than defined categories (e.g. 'strongly agree', 'disagree', etc.) have been used, does not make the variable numeric. However, since information about the type of the variable is for the user's reference only, and is not entered into the SPSS/PC+ system, any vari-ables which are ambiguous with regard to their categorical/numeric type may be indicated by a question mark. Decisions regarding which analyses are appropriate to that variable may need to be considered carefully at a later stage – perhaps when the variable has been examined in preliminary SPSS/PC+ analyses.

(iv) **Range of valid values**: As part of the process of assigning each variable to column(s) we have to consider the range of possible values that the variable might take. It is useful to include a note of this range within the codebook. For example, certain coding or typing errors (which give rise to data with values lying outside the valid range) can be identified in later stages of analysis. The inclusion of the valid range information is most useful for categorical variables.

(v) **Value labels**: For categorical variables, especially, it can be useful to include, within the SPSS/PC+ file which will be created, labels for some or all of the values which the variable may take. Thus for the variable *sex*, we might provide the label 'male' for the coding 1 and the label 'female' for the coding 2. Value labels are optional and their presence or absence does not affect the analyses performed. If labels are provided, however, SPSS/PC+ can include them in screen displays and printed output, and thus make the display of results more readily understandable by the user. When compiling a codebook for your project it is a good idea to work out a set of meaningful value labels.

Value labels may be provided for certain variables which are not clearly identifiable as either truly categorical or truly numeric in nature. Thus certain of the values obtained by using the traditional seven-point scale may be given labels (e.g. 1 = 'strong agreement', 7 = 'strong disagreement'). We see from this example that labels can be provided for only **some** of the values which are valid for a particular variable.

(vi) **Assigning missing values**: A special value should be assigned to each variable for coding missing data. In preparing the codebook you should assign one particular value (for example the number 9 or $\emptyset$, or perhaps 99 for a two-digit variable) as the missing value. The value assigned as the missing value should of course be one which cannot occur naturally in the data and which is therefore not within the valid range. If, during coding, missing or spoiled data is encountered, the missing value(s) should be entered into the columns assigned to the relevant variables. Begin coding again with values within the valid range when the data resume normally.

(vii) **Variable labels**: To make the output from SPSS/PC+ easily interpreted, variable labels, as well as value labels, may be included in the information provided to the system. The labels should be devised to provide maximum information, but should not be too long. Some thought should therefore be given to such labels when the codebook is compiled. The labels may be up to 60 characters in length and can contain spaces and various symbols. Variable labels are optional and their presence or absence has no effect on the analyses performed.

Using the examples (a) to (e) (pp. 18 and 19) as if they appeared in that order at the beginning of a single questionnaire (preceded only by the variable *id*, 'subject identification number') Table 2.2 illustrates the form which a codebook may take.

Conducting the survey

The next step is to produce multiple copies of the questionnaire and to distribute these to a suitable respondent population. Special care is needed with sampling, particularly if the survey is aimed at reaching a representative sample of a wider population (such as 'unmarried males' or 'working class wives'). Complex strategies are available for ensuring that samples are representative, and if your survey is of this type then you would do well

Table 2.2

Variable name	Col/s	Type	Values	Value labels	Miss.	Var. label
ID	1–3	N	–	–	999	Subject no.
EQRIT	4	C	1 to 7	Strongly agree Strongly disagree	9	Agree with equal rights
OCCUP	5	C	1 2 3 4	Manual Clerical Manager Profess.	9	Occupation – social class
FAVCOL	6	C	1 2 3 4 5 6 7	Red Orange Yellow Green Blue Indigo Violet	9	Favourite colour
TVHOURS	7–8	N	–		99	Number hours of TV/week
FAVGAME	9	C	1 2 3 ... 8	Hockey Football Skiing Chess	9	Preferred sport from 8 supplied
HOCKEY	10	C	1 2	Played Not played	9	Hockey in past month?
FOOTBALL	11	C	1 2	Played Not played	9	Football in past month?
SKIING	12	C	1 2	Played Not played	9	Skiing in past month?

to consult a specialist text. Initially, you might consult the **sampling procedures** entry in the glossary (Appendix G).

If the questionnaire is not to be completed by a captive population such as a class of students it may be necessary to arrange for pre-addressed envelopes to be included with the questionnaire, or pick-up points to be specified, in order to encourage a high rate of return. Unless the population is captive a 100 per cent return rate cannot be expected, and it is likely that the respondents who fail to return their questionnaires will be different in some important way from those who did, thus limiting the extent to which we can generalize from the analyses obtained.

2.4 Coding the data

When the completed questionnaires arrive you can begin to code the data using the codebook which has been compiled. The aim is to reduce all of the information in each questionnaire to a single string of digits for each case. Sometimes it will be possible to enter the information directly into a computer data file from the questionnaire. With more complex questionnaires, however, several stages may be necessary.

For example, a respondent's answers may first be coded onto the coding field of the questionnaire itself, and later transferred to special data sheets. Subsequently, the information from the data sheets can be entered into a computer file. Data sheets are simply empty grids of 80 columns by 25 rows with the columns numbered at the top of the page and with useful dividing lines marked after every 10th column. Using such a grid helps the coder to make sure that data relating to each variable are placed in the appropriate column(s). Because each subject has been given the same questionnaire a parallel set of answers must be obtained for each case. A value must be included (even if it is sometimes a missing value) for each respondent on each variable.

Thus the data matrix obtained as a result of the scoring process will take the following rectangular form:

```
0011191432421342132341411412412342142314234
0022223213131321323312313123123131313131222
0031234234333343424333333334225653455555444
0042193434444445454444453211244314232332344
0052202342342342342342342344456234234444233
0061234323423423432423424242424242342342344
```

In this case, each row across the page represents the set of data relating to one case. Sometimes all of the data relating to a single case cannot be accommodated within a single row and two or three rows (each row is sometimes referred to as a record) will be necessary for each subject.

If data for a single subject occupies two records, the data matrix may appear in the following form:

```
0011191432421342132341411412412342142314234676576571623561253331236
1212221111111122222211222
0022223213131321323312313123123131313131222334612543654352433462546
2221212222211122212112222
0031234234333343424333333334225653455555444342165436541654654654615
1212121212221222212222122222
0042193434444445454444453211244314232332344425425425435412354215421
2212222111122121212122221
0052202342342342342342342344456234234444423313243243212125556546465
2122212222212111112222111
0061234323423423432423424242424242342342344126565465525242654254652
122221212121222222211111
```

It can be seen that, although data from each subject is now accommodated over two lines, the rectangularity of the grid is in fact maintained.

Returning to the example of the shorter matrix shown earlier, we can see that the first respondent in the study (given the identification number 001) has supplied a set of answers which have been coded into the following row of numbers:

Subject 1   001119143242134213234141141241234214231423 4

The data from subjects 2 and 3, who produced somewhat different answers to the same set of questions, are given in the second and third rows:

Subject 2   002222321313132132331231312312313131311222
Subject 3   003123442343333434243333333342256534555555444

A single **column** (or, sometimes, two or three columns together) will always include the coded values for the answers given to a particular questionnaire item (or other variable).

Notice that the matrix is perfectly aligned on the right edge. A ragged right edge to the matrix, like this:

Row 1   001119143242134213234141141241234214231423 4
Row 2   002222321313132132331231312312313131311222
Row 3   00312344234333343424333333334225653455555444
Row 4   0042193444444545444445321124431423 2332344
Row 5   0052202323423424234234234445623423 44444233
Row 6   00612343234234234324234242424242342342344

would indicate that mistakes have been made in entering the data. Perfect alignment of columns throughout the data is essential because the computer will take all numbers in any column (or, sometimes, in a group of two or three columns together) as representing the same variable (e.g. answers to the same question) for all subjects.

Using our sample matrix again, we can number the columns 1 – 43.

```
      COLUMN              111111111122222222223333333333 4444
      NUMBERS   12345678901234567890123456789012345678 90123

(Subject 1)     001119143242134213234141141241234214231423 4
(Subject 2)     002222321313132132331231312312313131311222
(Subject 3)     00312344234333343424333333334225653455555444
(Subject 4)     0042193434444454544445321124431423 2332344
(Subject 5)     0052202342342342423423423444562342344444233
(Subject 6)     00612343234234234324234242424242342342344
```

Now let's examine in some detail just the first 10 columns of this dataset. For the sake of clarity, the next display includes spaces placed between the each pair of slices in the dataset (a slice corresponds to the data for a single variable):

```
COLUMN                                         1
NUMBERS              123  4  56  7  8  9  0

(Subject 1)          001  1  19  2  4  3  2
(Subject 2)          002  2  22  1  2  1  3
(Subject 3)          003  1  23  4  2  3  4
(Subject 4)          004  2  19  1  4  3  4
(Subject 5)          005  2  20  1  3  4  2
(Subject 6)          006  1  23  4  3  2  3
```

In this example the first three columns (**1, 2** and **3**) contain the respondent's identification number (the variable *id*). The use of three columns means that we can accommodate up to 999 subjects with separate *id*s. If we had a smaller respondent group (less than 100) we could make do with two columns (*id* values from 01 to 99), but since we may want to extend our survey later it might be best to keep the *id* format to permit up to 999 cases (larger surveys might require four or even five columns for the *id* number). Notice that the three-column example above requires that we give subject 1 an *id* code which extends over all three of the columns assigned for the *id* variable – that's why the coding for this subject is not 1 but 001.

Column 4 might contain the data for the variable *sex*. In this example, if 1 has been used as the code for males then we could see from the dataset that subjects 1, 3 and 6 are male and that subjects 2, 4 and 5 are female.

Columns 5–6 might represent the variable *age* – the age of the respondents in years. If the survey included only children under 10 then a single column might be assigned (or we might then want to specify age in months and use three columns). It is good practice to be generous in allowing column assignment, although over-generosity may lead to the tedious task of entering lots of unnecessary 0 values.

Columns 7, 8, 9 and **10** contain single-column variables. Whatever variables have been coded in these columns, it appears from our sample matrix that the responses of respondents 1 to 6 have all been coded within the range 1 to 4.

Let us suppose that the data in column **7** represent responses to the question '**How far do you agree that women should have the same rights as men?**', with 'strongly agree' coded as 1 and 'strongly disagree' coded as 5. We could name this variable *sper* and later provide it with a label 'support for equal rights'.

At this stage we can begin a cursory examination of the data. Notice that none of our subjects (out of our sample of only six) strongly disagreed with this statement about equal rights – there is no code 5. The two respondents who 'disagreed' with the statement (code 4) were men (coded 1 on 'sex', column **4**) and all three women 'strongly agreed' with the statement.

In drawing any conclusions from our data, however, we might wish to draw attention to the fact that this is a young adult sample. From the *age* columns (**5–6**) we can find that the average age overall is 21.0 years. Looking at the *age* columns in association with the *sex* column (**4**) we can calculate that the average age of the men (coded 1) is 21.66 years and that the average age for the women is 20.33 years.

This kind of cursory examination of the data (sometimes referred to as 'eyeballing') is interesting, and we've already begun to think in terms of:

frequencies	there are three men and three women
averages	the average age is 21
differences	the average age of the men is 21.66, whereas the average age of the women is 20.33; also, the women are more supportive of 'equal rights' than the men
associations	it looks as if the variable *sex* may be related to *sper* – support for equal rights

Such analyses lie at the heart of what SPSS/PC+ is designed to do (more accurately, less tediously, and with much larger datasets). Thus to explore such aspects of our data further, with a much larger sample and using more complex statistical procedures, we need to 'get into' the SPSS/PC+ system. SPSS/PC+ can handle many hundreds of cases, and any single dataset can include up to 200 variables (such as *sex*, *age* and *sper*). Furthermore, SPSS/PC+ will produce neat labelled tables which will enable us to examine and report our data thoroughly.

Now that you have worked through this chapter you should be able to:

- devise a structured questionnaire
- examine any questionnaire and analyze it in terms of the variables it contains
- devise a coding system for each variable
- transform completed questionnaires into a data matrix

This presentation of SPSS/PC+ focuses on data derived from questionnaires. The system can also be used to analyze data obtained from other kinds of study, including marketing projects, demographic studies and psychological experiments. In some cases there will be little need for an elaborate coding system. Each item of information collected from all subjects (or about each company, etc.) can be regarded as a separate variable. The rule for formatting the dataset so that single- or multiple-column slices contain data on a particular variable still holds. Information on the use of decimal data (e.g. 5.328, 66.23) and alphanumeric data (i.e. data coded in terms of letters instead of numbers) will be provided later (in Chapter 16).

3   SPSS/PC+ and the Computer

In this chapter we will begin to use the microcomputer to start our interaction with the SPSS/PC+ system.

SPSS/PC+ runs on an **IBM PC/XT** or **IBM PC/AT** (or on a machine which is a close compatible of one of these microcomputers). Such machines store information on a 10-, 20- or 40-megabyte hard disk. The hard disk is an integral part of the machine, capable of storing up to 40 million characters (numbers or letters). There will also be a slot at the front of the machine to accommodate **diskettes** – 5.25 inch floppy disks. This slot provides an opening to the diskette drive which is known as drive A (if your machine has two diskette drives, one will be labelled drive A and the other drive B).

Linked to the main computer module there will be a keyboard – a kind of typewriter which enables the user to communicate with the computer – and a display screen. A printer will also be linked to the machine. Together these units make up the hardware of the system. Computers run software or programs, and the SPSS/PC+ system is an example of a powerful program.

In your interaction with the microcomputer you will be writing, reading and using various types of **file**. Files can contain data, data descriptions, commands or any combination of these. Within files, information is stored in one of two ways – either in **binary form** (i.e. as **bits** of information) or as **text** (in ASCII form, i.e. as **alphanumeric** characters). The binary form is more efficient in some ways but ASCII files have the advantage that they can be easily amended or **edited**. Eventually we will learn how to instruct the machine to create binary files, but the first files we will create will be written to disk as ASCII (or text) files.

Because the hard disk is capable of storing many hundreds of different files, it is organized into different logical sections known as **directories**. There is a basic **root directory** from which other directories branch. Each branch may have several sub-branches, etc., so that the logical structure of the hard disk is rather like a tree. **Paths** can be made between one directory and others so that files stored within one directory can be accessed from others. If the system you are using has been set up as described in Appendix A, then the directory named SPSS (containing all of the SPSS/

PC+ system programs) will be accessible automatically from any other directory. To use the computer efficiently it is best to create your own directory and to work from this. Before describing how such a personal directory can be created, however, we need to understand the different levels of program which will be involved in our use of SPSS/PC+ on the microcomputer.

Using SPSS/PC+ involves dealing with three levels of program:

- **DOS**: the disk operating system; a control program
- SPSS/PC+ itself
- **REVIEW**: the SPSS/PC+ editor program

3.1 The three program levels

The first level is **DOS**, the disk operating system. This consists of a group of programs which enables the user to manipulate information stored either on the hard disk or on diskettes. **DOS** can be used to create, copy and edit files, to run programs, and to organize and switch between directories. **DOS** also handles communication between the computer and the various peripherals (for example, the printer).

The second level is SPSS/PC+ itself. This contains all of the data-handling and statistics programs of the SPSS/PC+ system.

The third level is **REVIEW**. This is an editor program within the SPSS/PC+ system which allows the user to create and edit files. **REVIEW** is SPSS/PC+'s own editor program. Other editor programs (such as the **DOS** editor EDLIN or the popular word-processing system WordStar) can also be used to create files for use by SPSS/PC+ (a general description of how SPSS/PC+ files can be created and edited using alternative programs is given in Appendix B). However, the fact that **REVIEW** is fully integrated with SPSS/PC+, and can be entered directly during an SPSS/PC+ session, means that it is generally the most efficient editor to use. It is recommended that you spend the little effort necessary to learn how to deal with **REVIEW** (a detailed presentation begins in the next chapter), even if you are already familiar with one of the other editors.

If **REVIEW** is being used as the editor program, a typical session will involve entry through **DOS** to SPSS/PC+, followed by a series of changes back and forth between **REVIEW** and SPSS/PC+.

The next section describes how you can move between the different levels of program: **DOS**, SPSS/PC+ and **REVIEW**.

3.2 Moving between program levels

Entering DOS

DOS is usually in operation as soon as you switch on the computer, so entering **DOS** usually involves simply switching on the machine. When internal checks have been made (this usually takes about one minute) the **DOS** prompt (**C>**) will automatically appear. **DOS** is now in operation, waiting for a command. The position at which you can type a command will be highlighted on the screen at the **cursor position**. You will (automatically) be in the **root directory**. If you have previously created a directory for yourself then you can enter your directory with a change directory (**cd**) command, followed by the appropriate directory name. After typing any instruction into the computer you should press <RETURN>. This **enters** the instruction into the system, e.g.

```
C>cd\johndoe    <RETURN>
```

If you have *not* previously created a directory for yourself you can create one with a make directory (**md**) command, followed by the name you wish to give to your directory. The name can be up to eight characters long. For instructions to **DOS**, as for instructions to SPSS/PC+, no distinction is made between upper case characters (capitals) and lower case characters. Thus, make a new directory, e.g.

```
C>md\samspade          <RETURN>
```

Now change directory (**cd**) to enter your newly created directory:

```
C>cd\samspade          <RETURN>
```

(If at any stage you need to return to the root directory this can be done by keying **cd** and then <RETURN>.)

Assuming that the system has already been set up appropriately for the easy use of SPSS/PC+ (if it has not, then see Appendix A), you will now be able to access the SPSS/PC+ system from your own directory by following the instructions below.

Entering SPSS/PC+

SPSS/PC+ will have been pre-loaded onto the hard disk of your machine (unless you yourself are implementing the system, in which case see Appendix A). Although the SPSS/PC+ package will be stored on the hard disk in a directory named "SPSS" it should be accessible from any other directory, including your own directory and the root directory.

Once **DOS** has been loaded and you have entered your own directory you are ready to enter SPSS/PC+. However, for the system to operate you will need to insert a special diskette (the SPSS/PC+ **Key Diskette**) into drive A. Without this key, access to SPSS/PC+ is prevented. The **Key Diskette** system prevents unauthorized copying and use of SPSS/PC+ because this diskette cannot be copied. It is therefore especially valuable and needs to be treated with the utmost care.

With the **Key Diskette** placed in drive A, to load the SPSS/PC+ system simply type 'spsspc' and press <RETURN>.

```
C> .spsspc     <RETURN>
```

The SPSS/PC+ logo will appear, followed by a prompt which, throughout each SPSS/PC+ session, will tell you that SPSS/PC+ is awaiting a command. The prompt looks like this:

```
SPSS/PC:
```

Entering REVIEW

To enter into the editor program **REVIEW**, simply reply to the SPSS/PC+ prompt by typing

```
review.     <RETURN>
```

The period (.) should be included. This is the terminator and is always used to signal that the SPSS/PC+ command is complete:

```
SPSS/PC: review.
```

You will now be presented with a screen divided into two horizontal halves, with various items of information written on the display. You are now in the editor program, **REVIEW**.

Getting back

For now, let us ignore the display. We can **return to SPSS/PC+** by exiting from the **REVIEW** program. This is achieved by pressing <aF10> (see note on p. xvi).

You will now be returned to SPSS/PC+, with the SPSS/PC: prompt displayed on the screen. SPSS/PC+ is awaiting your next command.

To exit from SPSS/PC+ we use the command word 'finish.' (again terminated with a period, .):

```
SPSS/PC: finish.
```

When you do this you will first be presented with the message **End of session. Please remember your KEY DISKETTE**. The **DOS** prompt (C>) will then re-appear. You are now back in **DOS** and in the directory from which you entered SPSS/PC+.

You have now learned how to move back and forth between the three levels of program you will use in a SPSS/PC+ session.

Interacting with SPSS/PC+

From **DOS**, get back into SPSS/PC+ once again. Type

```
spsspc   <RETURN>
```

and wait, after the logo display, for the SPSS/PC: prompt.

There are many different types of command which can normally be issued in response to this prompt, but since we have not yet created any files containing information to be analyzed we are somewhat restricted at this stage. We can, however, explore some **operation commands** which allow the user to obtain useful on-screen information and to change some of the characteristics of the system.

Three such commands will be examined at this stage – **HELP**, **SHOW** and **SET**.

The HELP command

The **HELP** command is used to obtain on-screen information about a wide range of SPSS/PC+ topics.

(a) The command **HELP.** (without further specification) produces a screen display of general instructions on how to use the **HELP** facility:

```
SPSS/PC: help.
```

In response to the <RETURN> following this command, a **MORE** message will appear in the top right corner of the screen and a high-pitched beep will be heard. This is a common SPSS/PC+ response. It means that there is a further page of information to be presented. To produce a display of the next page you simply have to press **any** key once.

(b) The command **HELP all.** produces a screen display of the major **HELP** topics (e.g. files, case selection, etc.):

```
SPSS/PC: help all.
```

(c) The command **HELP** followed by one of the major **HELP** topics (such as files) provides either general information on the topic or a list of more specific topics on which **HELP** may be obtained, e.g.

```
SPSS/PC: help files.
```

(Sooner or later you will make a mistake in the syntax of one of the commands you issue to SPSS/PC+. The system will inform you of this, generally by displaying an error message with an explanation of your mistake. This is one of the most useful features of the SPSS/PC+ system. A very common error is to omit the terminal period (.) of a command. This will not produce an error message, but SPSS/PC+ will present a continuation prompt (:). If you press <RETURN> in response to this, SPSS/PC+ will then accept the previously issued command.)

(d) In later chapters we will examine many types of command which can be issued to SPSS/PC+, including **procedure commands**. These instruct the system to read data and often request statistical analyses. For many procedures the user may request certain **OPTIONS** and additional **STATISTICS**. One simple procedure is **MEANS** – this supplies certain basic statistics about groups of cases we have selected from our dataset. The command **HELP** followed by the name of a procedure and the word statistics, will provide the user with information about the additional **STATISTICS** available for that procedure, e.g.

```
SPSS/PC: help means statistics.
```

The **HELP** facility can also be used to gain help with files and the SPSS/PC+ syntax rules as well as several other elements within SPSS/PC+. If difficulties do arise during a SPSS/PC+ session then an appropriate **HELP** request may provide the information you need to solve your problem.

The SHOW and SET commands

The other **operation commands** to be discussed in this chapter are **SHOW** and **SET**. The **SHOW** command produces a display of the current state of many aspects of the SPSS/PC+ system. The **SET** command allows these aspects to be changed.

The **SHOW** display is produced merely by entering 'show' in response to the SPSS/PC+ prompt:

```
SPSS/PC: show.
```

No further specifications are necessary for or applicable to the **SHOW** command.

Much of the information displayed on the screen in response to the **SHOW** command can be ignored at this point, but let us look at some of the items and consider how certain aspects of the general operation of the SPSS/PC+ system can be changed by an appropriate use of the **SET** command.

Look, in particular, at two values which represent the current **page size** – **WIDTH** and **LENGTH**. Look also at the current on/off status of the **printer** and three other aspects – **MORE, BEEP** and **EJECT**.

These (and the other items included in the **SHOW** display) can be changed using various **SET** commands in response to the SPSS/PC+ prompt.

Thus, if the printer is currently off, it may be switched on with the command:

```
SPSS/PC: set printer on.
```

Or, if the printer is on, and you wish to switch it **off**, you would use:

```
SPSS/PC: set printer off.
```

These commands of course refer specifically to the interaction of the SPSS/PC+ system with the printer. Obviously the on/off switch on the printer itself must be on for the machine to operate at all.

When discussing the use of **HELP**, above, it was mentioned that when a further page of information was waiting to be brought onto the screen, SPSS/PC+ would issue a **MORE** message at the top of the screen and that a high-pitched beep sound would be heard.

The function of the **MORE** message is to allow the user to read one screen of output before the next one is presented. The SPSS/PC+ action is thus paused until, by pressing a key, the user brings the next page of information to the screen.

Sometimes, however, the user does not wish to read the display when it appears on the screen. S/he would prefer, perhaps, to read a printed form of the complete output at a later time. With the **MORE** function **on**, the user would need to wait at the machine and repeatedly press a key to move the output (both screen and print) on to the next page.

MORE can be switched **off** by using the **SET** command:

```
SPSS/PC: set more off.
```

and can be switched on again:

```
SPSS/PC: set more on.
```

The audio signal beep has two functions. A high-pitched beep merely alerts the user to the fact that the current display is being held and that a key-press is necessary if further output is to be displayed. A low-pitched beep signals an error.

If the user wishes to switch the beep **off**, this can be done by issuing the following **SET** command:

```
SPSS/PC: set beep off.
```

The width and length of the display page (as it appears both on the screen and in any printout from the printer) can be seen to be set at certain values. If the paper you are using is not of the appropriate size then these values can be changed. Although you may tailor your page-size by specifying exact values for the width and length, the most frequent and useful change sets the width to the standard **wide** paper (130 columns across) instead of the standard **narrow** paper (79 columns across). This can be achieved with the command:

```
SPSS/PC: set width wide.
```

Finally, you may use **SET** to change **EJECT** from **off** to **on**. It is often the case that a page of output from SPSS/PC+ occupies less than the full length of a page of paper. Normally, with **EJECT off**, the output is printed continuously so that if a page of output is one-half of a sheet of paper in length, then the full printed output of 10 pages will occupy only 5 sheets. This saves paper.

Sometimes, however, the user might wish for each page of SPSS/PC+ output be printed on a new sheet of paper. One advantage of this will be that no line of the printout will ever run across the perforations of the continuous paper which is generally used.

To specify that a new output page should begin on a new sheet of paper a **SET** command can be used to switch the 'EJECT' feature **on**:

```
SPSS/PC: set eject on.
```

As you are introduced to new features of the SPSS/PC+ system you will come to understand more of the items in the initially bewildering display elicited by the **SHOW** command. You will also learn when it is appropriate to use a **SET** command to change particular aspects of the system. Generally, it is not essential to change **any** aspect in order for SPSS/PC+ to run successfully because the settings which are provided automatically allow the system to function perfectly well (you may, however, have to **SET** the printer on). The automatic settings (including the names provided for certain types of file, etc.) provided by the system are known as *default* settings. They operate by default i.e., unless the user deliberately changes one or more of them with **SET** commands. A clever use of **SET** commands, however, can add flexibility and allow the user to work the system at maximum efficiency.

Having now learned how to switch between levels of program (**DOS**, SPSS/PC+ and **REVIEW**) and how to interact with SPSS/PC+ directly, using the **operation commands HELP, SHOW** and **SET**, it is time to consider how **files** can be created and stored on disk. We will start, in the next chapter, by using **REVIEW** to enter a dataset into a **data file** which can be written to the disk for later analysis by SPSS/PC+.

4 Creating a Data File

In this chapter we will present some basic guidelines for using the SPSS/
PC+ editor program **REVIEW**. We will use this editor to enter data and to
create a **data file** which can be stored on disk.

4.1 Using REVIEW

REVIEW is the SPSS/PC+ editor program. Because of its power and
flexibility, **REVIEW** is somewhat complex, and in this chapter only some of
the many features of **REVIEW** will be introduced. Enough information will
be given, however, to allow you to enter data which has been coded onto
questionnaire forms (or onto **data sheets**) into a **data file**. More details
about the use of **REVIEW** will be given in Chapter 12.

If you are still in **DOS** (i.e. if the screen presents you with the **DOS**
prompt **C>**) then type **spsspc** (and <RETURN>). The screen should
display the SPSS/PC+ logo, followed by the SPSS/PC+ prompt (SPSS/
PC:)

Respond to the SPSS/PC+ prompt by typing **review** followed by the
name of the data file **to be created**. File names often include an extension of
three letters following a period. Thus a word-processing file which contains
text might be given the name ".essay.txt", where ".txt" is the extension. It
is useful to think of files as having a 'forename' which comes before a
period (.) and a 'surname' which comes after. The forename can be up to
eight characters (letters and/or numbers) long. The three-letter surname
can be chosen to best describe the nature of the file. Thus a data file may be
given the surname ".dat". Throughout this book, files which contain only
data will be given names with this extension (thus "trial.dat", "gender-
.dat", etc). Thus we could use the **REVIEW** program to create a data file
by entering the instruction:

```
SPSS/PC: review "trial.dat".  <RETURN>
```

Notice that because "trial.dat" is the **name** of a file (rather than a direct
command like **HELP** or **REVIEW**, it is typed within quotation marks. The
final period (.) is used to signal the termination of the **REVIEW** command.

If the command has been typed correctly a screen will be presented on which we may start to enter data. We simply enter the numbers in our data matrix along the screen in a long string, without spaces. The data for each respondent should end with a <RETURN>, thus creating a new line on which to begin entering the data for the next respondent. When all of the data has been entered it will be saved on the disk in a file which will later be used by SPSS/PC+.

The screen will already display certain items of information. At the bottom, for example, there may be two messages – 'Num' and 'Ins'. If 'Num' is present this tells us that the bank of keys on the **number pad** (usually on the right-hand side of the keyboard) is switched to number mode. These keys can also be set to the alternative function mode. Look for a key labelled 'Func Lock', and press it once. The 'Num' message will disappear from the bottom of the screen. The number pad is now set to the function mode. Looking at the keys you will see that some of them, at least, have both numbers *and* short function labels (<PG DN> for 'page down', <DEL> for 'delete', etc.). With the 'Num' message absent from the screen the function mode is in operation, and depressing the keys will produce various function effects rather than causing a number to be written on the screen.

With the function mode in effect, press <INS>. Press it once and the 'Ins' message on the screen will disappear. Press it once again and the 'Ins' message will reappear.

When the 'Ins' message is present we know that the editor program is in **insert** mode. Sometimes it is useful to have **insert on** and sometimes it is more useful to have **insert off**. If the insert is **on**, any typed characters will be inserted at the current cursor point. Thus if we have a display:

    11113333

and we place the cursor at the first 3, then typing the figure 2 four times will produce the result:

    111122223333

With the insert **off**, however, the typed characters will **overwrite** any existing characters. The same key presses would thus result in:

    11112222

If a data item has been mistakenly omitted from a data line, therefore, we could insert the missing character(s) at the correct point with the insert **on**. If, however, we need to overwrite one or more incorrect numbers then we would do this with the insert switched **off**. With insert **on**, pressing <RETURN> will create a new line below the current line. With the number pad in **function mode**, one key will act as a **delete** key. Pressing this will delete the character at the current cursor position.

Before exploring the use of function keys (see the notes on the use of function keys on p. xvi of the Preface) we need to enter some data (it can be real data from a study of your own, random dummy data, or the data from the gender attitudes survey reproduced in Appendix E). If the data from a single case extends over more than one 80-column line, press <RETURN> at the end of the first line and continue entering the data for that respondent on column 1 of a second line. When all the data have been entered for the case, press <RETURN> and start entering the **next** respondent's data on a **new** line.

Continue for a number of subjects. If you notice at some point that the number you have just typed is a mistake, place the cursor over the mistaken number and press <DEL>. If you notice a mistake when you have already entered further data then move the cursor to the appropriate position (taking care that **Insert** is **on** or **off**, as appropriate) and insert any missing numbers, or overwrite or delete errors.

When you have become used to this data entry procedure you can start to explore the further editing facilities of **REVIEW**. After you have started entering a line of data press <⌃ F4>. You will see that this operates the line delete function and that the whole of that line of data has disappeared. Start entering the line of data again from column 1.

Next, deliberately leave out one subject's data. After, say, subject 15, enter the data for subject 17. When this is complete you can go back to insert the line or lines of data for subject 16. To do this return the cursor to the last line of data for subject 15 (at any point on the line) and press <F4>. You will see that a free line appears below the current line. You can now enter the data for subject 16.

To explore further the various functions which are available, press <F1>. You will be presented with a display of the various **REVIEW** commands. There are rather a lot. Don't worry about all of the available functions at this stage – many are merely short cuts to effects which can be achieved with a much more limited range of **REVIEW** commands. Notice the <F4> and the <⌃ F4> which you have already used. Now press <SPACE> to return to your data entry.

When you have entered all the data you wish to enter at this stage, correcting as necessary, you can store the data as a file on the disk. This will later be used for SPSS/PC+ analysis.

There are two ways of saving the data as a file. You can either save the file or make a block within the file and save the block.

To save the file

To save the complete contents of the file you have created, you use the command < ↑ F9>. This instructs the program to write the file to disk. You will be presented with a message which asks you to name the file to be stored and you respond to this by typing a suitable name (you do **not** need quotation marks, and you do not have to use the name you provided when initially creating the file). Press <RETURN> and the machine will store the data as a file on the hard disk. It will have the name which you have just provided.

To make a block and save it

Rather than saving the complete contents of a file you may sometimes need to save only a part of what you have written. To do this you can make a **block** of just some of the lines (as many as you wish) and save that. Thus within a dataset, you can make a block of some or all of the data you have entered, and then write the block to disk. The contents of this block will be stored as a file which can be retrieved for later editing or analysis. To create the block, you need to move the cursor to any position on the first or last line of the data you wish to save and then press <F7>. You will notice that the line begins to flash and that a message appears: **Waiting for second block marker**. Now move the cursor to any position on the line which forms the other end of the data block (i.e. the last line, if the line already marked was the first line of the block) and press <F7> again.

The block has now been marked and the lines on the screen which are included in the block will change to a different shade (or colour). A message will tell you that the block has been marked and also tell you how many lines are included in the block (**Block marked N lines**). To save a copy of the block on disk you press <F9> and the computer will ask for the name of the file to be written. Enter some suitable name (e.g. **trial.dat** or **gender.dat**, without quotation marks) and press <RETURN>. The block will now be written as a file on the hard disk, with the name you have chosen.

To exit from REVIEW

Having successfully saved a complete **REVIEW** file, or a block, you can exit from **REVIEW** by issuing the command <aF1∅>. You will be presented with the SPSS/PC: prompt.

Congratulations – you have created a data file which is now stored in your directory of the hard disk. At this stage you might like to obtain a printout of the contents of the data file you have created.

To do this you return to the disk operating system (**DOS**). Enter **finish.** to end the SPSS/PC+ session and return to **DOS**.

```
SPSS/PC: finish.
```

You will receive the **End of session** message, followed by the **DOS** prompt (**C>**).

Respond by asking the machine to print the file. The command takes the following form (make sure you use your own filename):

```
C> type trial.dat>prn   <RETURN>
```

If the printer is switched on you should receive a printout of your data. This can be examined for errors. It is worth checking that your data block is perfectly rectangular. If there is a wavy right edge then this indicates that you have either inserted or left out data digits (thus putting the rest of the line out of true). Such errors will be easy to correct when you next **REVIEW** the file.

Creating a data file with **REVIEW** is an important step towards using SPSS/PC+, but so far the file is just a long list of numbers. These must now be identified to the system as **variables**. The variables must be named, and the position they occupy within the data file must be fully specified. Additionally, the system should be informed of which values signify missing data, and labels may be supplied for particular variables and the values associated with them.

Such information is generally specified in a **data definition file**. Such a file is also created using an editor program. In Chapter 5 we examine how a data definition file may be created using **REVIEW**.

Table 4.1 summarizes the **REVIEW** commands encountered so far.

Table 4.1

Command	Key(s)
REVIEW help (This is the most useful REVIEW command. It produces a screen display of all the commands used by the REVIEW program – it can be called up at any stage and the screen display will not interfere with the file you are editing. You can return to continue editing by pressing <SPACE>.)	<F1>
Delete a line	<^F4>
Insert a line after current cursor position	<F4>
Write file to disk	< ↑ F9>
Mark block (beginning AND end)	<F7>
Write block to disk	<F9>
EXIT from REVIEW	<aF10>

5 Creating a Data Definition File

Creating a data file, containing only the data, is just a first stage towards using SPSS/PC+. The data must be **defined** (and is also usually more fully **described**) before it can be used. The data is defined in a **data definition file**. This usually contains the following types of command:

> **DATA LIST**: a list which gives the names of all of the variables represented by the data, in the order that they occur, and information about which column(s) each variable occupies
>
> **VARIABLE LABELS**: (optional) variable **labels** should not be confused with the variable **names** specified on the DATA LIST
>
> **VALUE LABELS**: (optional) value labels for some or all of the values for each variable
>
> **MISSING VALUE**: information about which value, if any, represents missing data for each variable

There are few rules in SPSS/PC+ about the order in which the above commands are entered, but there is one rule which is very important. A variable must be declared in a **DATA LIST** command **before** any labels are provided for it or a missing value specified for it.

Together, all of this information is said to constitute the **data dictionary**. Data definition files contain either the data dictionary only, or both the data dictionary and the data itself.

Although data dictionary information can be entered into SPSS/PC+ interactively (i.e. the commands can be entered directly in response to SPSS/PC: prompts) it is often more efficient to create permanent **data definition files**. Such files can be created using **REVIEW** (or another editor such as EDLIN or WordStar – if you intend using an editor program other than **REVIEW** then consult Appendix B).

We will use **REVIEW** to create a simple SPSS/PC+ data definition file which can be used with the data which has been saved in the data file (see Chapter 4). We will also use **REVIEW** to re-edit the file if any errors emerge, and to correct any data-entry errors which have been detected in the printout.

To enter **REVIEW**, we first enter SPSS/PC+. We do this simply by responding to the **DOS** prompt with 'spsspc':

```
C> spsspc   <RETURN>
```

The SPSS/PC+ logo will appear, followed by the prompt, and we then issue a **REVIEW** command which includes an appropriate name for our data definition file. It is a good idea to use the extension (or 'surname') "def" for data definition files, e.g.

```
SPSS/PC: review "gender.def".
```

The system will respond by presenting a screen on which we can start to enter information. We will enter a **DATA LIST**, followed by **VARIABLE LABELS**, then our **VALUE LABELS**, and then the **MISSING VALUE** information. Data can be added directly to a data definition file, or an instruction may be included which will enable the system to read the relevant data from an existing data file.

Information about each variable in the dataset should be readily available in the codebook which will have been devised before the data was coded into number form. This information will help us to write the series of commands which will form the data definition file. The codebook for the gender attitudes survey which provides a working example throughout this book is reproduced as Appendix D. A codebook usually contains the variable names (in the order that they occur within the dataset), the number(s) of the column(s) in which each variable occurs, the missing value assigned to each variable, and any variable and value labels.

The variables must be **defined** by a **DATA LIST** command **before** they are labelled, etc. The other commands listed above can be entered in any order and there can be multiple **MISSING VALUE**, **VARIABLE LABELS** and **VALUE LABELS** commands within a single file.

5.1 The DATA LIST command

For data to be usable by SPSS/PC+ they must be defined. Each variable has to be named and the program has to be informed (in a formatting instruction) which columns of the dataset each variable occupies. **Variable names** and **formatting instructions** are specified in a **DATA LIST**. For example,

```
DATA LIST /ID 1-4 Sex 5 Age 6-7.
```

Thus each variable is named (*id*, *sex*, *age*) in the order in which the relevant data appear in the dataset, and after each name the column(s) occupied by the variable are fully specified. In the example above, data for the variable *id* occupy columns 1 to 4 in the data matrix, and the variable *sex* is coded in column 5. For the **DATA LIST**, and for all other SPSS/PC+ commands, no distinction is made between upper-case and lower-case characters.

SPSS/PC+ is generally tolerant of extra spaces added into commands as long as they are between (rather than within) elements. Also, there must be at least one space between the variable name and the column number (otherwise the column number would be interpreted as part of the variable name).

Variable names

Variable names consist of up to eight characters (letters or numbers) starting with a letter. Names can be individual specifications (like *sex* and *age*), or may be of the type *item14*, *v1*, *var65* (for item 14, variable 1 and variable 65).

On the data list (and on most other commands) variables can be listed using the **TO** convention. Thus:

```
VAR001 TO VAR008        will create 8 variables
V1 TO V105              will create 105 variables
```

But note that *v1* is treated as a **different** name to *v001*. Thus, after creating a variable *v1* or *v001* (etc.), the exact original form must be used when referring to that variable in any later command.

Formatting instructions

The instruction specifying the column(s) in which a particular variable is to be found is referred to as a formatting instruction. In the example

```
DATA LIST /ID 1-4 Sex 5 Age 6-7.
```

we can see that columns are specified by a single number or by a number range (e.g. 1–4). There must be at least one space between the name and the column number(s) and between the column number(s) and the next variable name.

Where the **TO** convention is used, the columns occupied by the group of variables are assumed to be equally divided between the variables. Thus the specification:

```
VAR001 TO VAR031 10-71
```

indicates that the 31 variables occupy **two columns each** (notice that the variable and column numbers are inclusive – VAR004 TO VAR014 specifies 11 variables; a column specification 35–46 includes 12 columns).

The following example provides the formatting information for 10 single-column (or single-digit) variables, followed by one three-column variable, followed by **four** two-column variables:

```
VAR001 TO VAR010 10-19 VAR011 20-22 VAR012 TO VAR015 23-30
```

If the set of data for a single case occupies more than 80 columns, some of the data for that case will have been entered on an additional line (or perhaps over several additional lines). The data list must specify where, in the list of variables, new data lines begin. Column numbering for a second (or third) line then begins from 1 again (**not** 81 or 161). A new line is indicated by a slash (/) and the system then interprets a 1 as referring to the first column of the second (or third) line.

Remember that the data for a single variable (for each case) must be **wholly** placed on one line – a variable cannot begin on one line and end on the next. Also, only variable names applicable to data on one line can be specified before the slash. The naming of further variables must start again when a slash has indicated a new data line.

Thus the following specification is *not* permitted:

```
ID 1-4 VAR001 TO VAR080 5-80 / 1-4 Age 5-6
```

The legitimate form of this command is:

```
ID 1-4 VAR001 TO VAR076 5-80 / VAR077 TO VAR080 1-4 Age 5-6
```

This tells the system that after reading the ID and variables 1 to 76 from the first line, it should read variables *var77* to *var80*, and the variable *age* **for the same subject**, from columns 1-6 of the next line.

Following all of these rules, a valid **DATA LIST** command might look like this:

```
DATA LIST /ID 1-4 Sex 5 Age 6-7 VAR001 TO VAR006 8-13 VAR007 14-16
VAR008 TO VAR039 17-80 / VAR040 TO VAR047 1-16 Q1 17 Q2 18-19
Q3 TO Q12 20-29.
```

Here we can see that the **DATA LIST** itself may run over 80 columns. At a convenient point (not, for example, in the middle of a variable name or column specification) you may press <RETURN> and continue entering the information on the next line. Additional blank lines are not permitted **within** a command (although they may be inserted **between** different commands in the file). The **DATA LIST** command must end, like other commands, with a period (.).

You should now be able to enter your own **DATA LIST** command. Correct any typing mistakes as you go. The **REVIEW** editing functions operate in precisely the same way as for data entry (press <F1> for a display).

5.2 The VARIABLE LABELS command

Extended descriptions (labels) can now be provided for some or all of the variables. The labelling of variables (and also of values) is optional and the presence or absence of labels does not affect statistical analyses or any other data manipulation by SPSS/PC+. If they have been provided, however, labels are used to annotate displays of results, including tables, and they can thus make SPSS/PC+ readily interpretable by reminding the user of what the variables and their values represent.

Variable labels can be up to 60 characters in length and can include spaces and any other characters. It is best, however to keep them as succinct as possible, especially since the system sometimes prints long labels in truncated (shortened) form. Otherwise, the system prints the labels exactly as they have been specified. This means that any spelling errors, and any use of upper-case and lower-case characters in the original specification of the labels, will be accurately reflected in the output.

Variable labels are entered by specifying the name of the variable and then, **within quotation marks**, the label to be given to it. Consecutive specifications are separated by a slash (/). For example:

```
VARIABLE LABELS ID "Identification" / SEX "Sex of subject"
/Age     "Age of subject"
/VAR001 "Attitude to women's work"
/VAR002 "Attitude to women's education"
/VAR003 "Attitude to child-rearing"
etc.
```

Although two or more variable labels can be specified on a single line it is often preferable to give a separate line to each variable, as shown above. You cannot begin a label on one line and finish it on the next. Additional blank lines are not permitted **within** the command, and the final variable label specification should end with a period (.).

In some cases it may be appropriate to provide exactly the same label for more than one variable. In this case, rather than writing the same label a number of times, we could enter a variable **list** and supply the common label just once. Thus:

```
/VAR001 VAR003 VAR004 VAR016 Q5 Q7 "Employment attitudes"
```

Here the six variables named would all take the common label 'employment attitudes'. Such lists can also be used within many other commands (for example when instructing the computer to carry out the same statistical analysis for several variables).

When appropriate, the **TO** convention may be used. Thus:

```
/VAR001 TO VAR016 "Attitudes to women"
```

Here 16 variables would each take the label 'attitudes to women'.

The **TO** convention can be used to specify any group of successive variables in the dataset whether or not they have the same type of name. Thus VAR1 TO TIME6 would be valid if it was used to refer to a list of successive variables. The **TO** convention is very useful when a number of successive variables are all to be treated in the same way (in this case all given the same variable label). Thus:

```
/VAR004 TO Q5 "Attitudes to men"
```

would specify that all of the variables listed on the original data list between *var004* and (up to and including) *q5* would be given the common variable label 'attitudes to men'.

Remember that variable labels are optional. You can label as many or as few variables as you wish.

5.3 The VALUE LABELS command

A display which includes only the coded values for a variable – 1, 2, 3, etc. – is often inconvenient, for the user will frequently need to be reminded of what particular values represent. The display or printout (including the output from statistical analyses, and tables and graphs) will be considerably more meaningful when labels have been provided for the various values.

Value labels are optional and their presence or absence does not affect the analysis. Where the data for a variable are truly numerical (as for *age* recorded in terms of years), rather than codes, the **variable** label will generally be sufficient to remind the user of the nature of particular values. In such circumstances value labels would serve no useful purpose. Labelling is particularly useful, however, when data have been coded according to a category system. Value labels are specified on a **VALUE LABELS** command and a single command is sufficient for providing labels for many variables (although it is also permitted to use several **VALUE LABELS** commands for different variables in the dataset).

If different variables have the same value labels you may provide a single set of value label specifications for a list of these variables. In listing these variables the **TO** convention is permitted. Thus:

```
VALUE LABELS Sex 1 "Male" 2 "Female" /VAR001 1 "Agree" 7 "Disagree"
/VAR002 TO VAR005  1 "Positive"  4 "Neither"  7 "Negative"
/VAR006 VAR008 VAR012 1 "Very much" 2 "A little" 3 "Not at all"
etc.
```

Here the variable *sex* has been given the value label 'male' for the value 1 and the value label 'female' for the value 2. *Var001* has been provided with value labels for just two of its values – 1 (labelled 'agree') and 7 (labelled 'disagree'). The next four variables (*var002* to *var005*) have each been given the same set of value labels for three values (1, 4 and 7) and three further variables have also been given a set of value labels.

Each set of labels applying to a single variable or to a list of variables must be separated by a slash. Although a set of value labels can be split over two lines no single label can be split; thus any one label must begin and end on the same line. Blank lines must not be inserted within a **VALUE LABELS** command, and the final value label specification should end with a period (.).

Notice, in the example above, that not all the values associated with a particular variable need to be labelled. Thus for a seven-point scale of agreement or disagreement, for example, labels may be provided only for the extremes:

```
/VAR002 TO VAR005  1 "Strongly agree"  7 "Strongly disagree"
```

5.4 The MISSING VALUES command

A missing value is a code assigned to data which is missing from the dataset (for example, where a respondent has failed to answer a questionnaire item). It is a number which is outside the range of valid values of data for that variable. SPSS/PC+ will recognize that data coded with this value is missing and will not include the value in, for example, the calculation of an average for the variable. The system can only do this, of course, if it has been informed that a particular value for the variable signifies missing data. Such information is provided to the system on one or more **MISSING VALUE** commands. In such a command a variable name (or a list of such names) is followed by the appropriate missing value code in brackets:

```
MISSING VALUES Sex (9) / Age (99) / VAR001 TO VAR062 (0)
/VAR073 TO VAR080 (9).
```

This instructs the system that 9 is the missing code for *sex* and also for the variables *var073* TO *var080*, that 0 is the missing code for *var001* TO *var062* and that 99 is the missing code for *age*. No missing value has been provided for some variables which may be presumed to be in the dataset (*var063* to *var072*). Although it is not mandatory to provide missing values for all variables (or indeed for any variable) it is certainly good practice to do so. It should always be presumed that there will be missing data.

As well as lists (including the **TO** convention) the keyword **ALL** can be used:

```
MISSING VALUES ALL (0).
```

This assigns Ø as the missing value for each variable in the data list.

We have now used **REVIEW** to write a data definition file which contains the **DATA LIST**, the **MISSING VALUE** specification and the (optional) **VARIABLE LABELS** and **VALUE LABELS** commands. We could save this as a data definition file except for one important thing – we have not yet included either the data or an instruction specifying a file in which the data may be found.

5.5 Inserting data

Data can be directly inserted into the data definition file or, alternatively, a separate data file can be referred to in the **DATA LIST** command. In the latter case the data from the data file are brought in each time the **DATA LIST** command is executed.

To insert data directly

If we wish to insert data directly into a data definition file (such data are referred to as in-line data) we can either key in the data as described in the instructions for creating a data file (Chapter 4) or we can use a **REVIEW** command to insert a previously created data file. If the latter strategy is used, the dataset from the original data file becomes an integral part of the data definition file.

Any file which has in-line data must include a **BEGIN DATA.** command immediately before the first line of data:

```
BEGIN DATA.
```

(Note that the period (.) **must** be included.) This signals that data follows.

When this command has been written we either begin typing in our data for each subject, as described in Chapter 4, or we can insert a previously created data file. To do this we need to press <F3>. The computer will ask for the name of the file to be inserted – **Name of file to be inserted** – and we should respond with the name of our data file (e.g. "gender.dat") *without* quotation marks.

Data from the data file will be read into the data definition file being created and will appear on the screen. This is a good time to make corrections to any errors which a printout may have revealed.

When all the data have been included (either by direct data entry or by inserting a data file) we enter an **END DATA.** command on a new line immediately following the last data line:

```
END DATA.
```

(Again, the period **must** be included.) We have now created a data definition file containing in-line data and information which defines the data. This file can be saved by writing the file to disk (see below).

To specify a data file

If we choose not to include the data as an integral part of the data definition file, we can include, in the **DATA LIST** command, a reference to a data file already stored on disk. The data from that file will then be read automatically whenever the the **DATA LIST** command is executed. To amend the above **DATA LIST** command in order to refer to the previously created data file (e.g. "gender.dat") we would insert **File=** and the name of the file to be used:

```
DATA LIST FILE= "gender.dat" /ID 1-4 Sex 5 Age 6-7
VAR001 TO VAR006 8-13 VAR007 14-16  VAR008 TO VAR039 17-80
/VAR040 TO VAR047 1-16 Q1 17 Q2 18-19 Q3 TO Q12 20-29.
```

Since there is no actual data in the file we do not include either a **BEGIN DATA.** or an **END DATA.** command. Each time the **DATA LIST** command is executed, data is read from the current version of the file with the specified name.

To write the data definition file to disk

A data definition file is saved (written to the disk) in the same way as a data file, by using either the file save or block save **REVIEW** commands introduced in Chapter 4.

To save the whole file, press <↑F9>, name the file and press <RETURN>. The data definition file (e.g. "gender.def") will now be stored on disk.

We can now exit from **REVIEW** by pressing <aF10>.

5.6 Testing the data definition file

When the data definition file has been written to the disk, it is a good idea to test it immediately to see whether it is acceptable to SPSS/PC+. In the early stages of using SPSS/PC+ there are likely to be one or two errors (missing periods (.) for example, or misspelt variable names on the labelling commands).

It often proves useful, when SPSS/PC+ diagnoses an error, to refer to a printed copy of the data definition file and to examine this while reading the SPSS/PC+ error message displayed on the screen.

To obtain such a reference printout of the data definition file we need to get back to **DOS** (exiting from SPSS/PC+ by entering 'finish.' in response to the SPSS/PC: prompt). We then issue a **DOS** print command, e.g.

```
C> type gender.def>prn
```

Next, get back into the SPSS/PC+ system (by entering 'spsspc' in response to the **C>** prompt).

When you are presented with the SPSS/PC: prompt you can attempt to **INCLUDE** the data definition file:

```
SPSS/PC: include "gender.def".
```

SPSS/PC+ displays each command line as it is processed and performs various tests on each line as it is included. If a problem is diagnosed an error or warning message will be displayed and a low-pitched beep will be heard.

The system checks to see whether variables specified on later commands have been declared on the **DATA LIST** command. Thus if the spelling of a variable name on a **VARIABLE LABEL** specification is different from that on the **DATA LIST** command, this will produce an error message. SPSS/PC+ will claim that you are attempting to label a variable which does not exist.

In addition to error messages, SPSS/PC+ also issues warnings. These will not prevent the file from being **included** but you should study them carefully and try to correct the problem. Study any error message or warning in conjunction with both the line displayed on the screen and your printout of the file contents. The printout can be amended to include the corrections which will need to be made when you re-edit the file.

SPSS/PC+ will detect only certain types of error. Among the common errors which SPSS/PC+ will diagnose are:

- a terminator period omitted so that the next line is treated as a continuation
- errors in format specification on the **DATA LIST**
- misspellings of variable names, so that a variable which appears in a ·labelling command has not been declared on the data list
- blank lines inserted **within** a multi-line command
- variable labels or value labels split across two lines
- illegal characters within the dataset – sometimes a letter will have been entered in error (it is particularly likely that a letter O may have been typed instead of 0)
- exceeding limitations – each procedure has limits (for example, on the number of variables that can be included in a command, or the number of computations which can be requested)

SPSS/PC+ will *not* detect:

- numerical errors in data entry
- spelling errors within quotation marks
- labels which have been omitted
- errors in the labelling of values or variables

Sometimes SPSS/PC+ will be over-enthusiastic in spotting errors. Thus if a **DATA LIST** command contains an error, no valid variables will have been declared to the system and any subsequent reference to any variables on the **DATA LIST** will be diagnosed as erroneous. A subsequent correction of the **DATA LIST** command may therefore prove sufficient to correct a whole series of dependent errors diagnosed by the system.

When errors have been diagnosed by SPSS/PC+ you should **REVIEW** the data definition file, edit out the errors and save the file. To re-edit a file first call up the file with **REVIEW**, citing the name of the file to be edited:

```
SPSS/PC: review "gender.def".
```

After the necessary changes have been made (by moving the cursor to the appropriate places and using the normal **REVIEW** editing functions) you can save the file with $< \uparrow F9>$. Once an appropriate name has been established for the file, the original name should be retained for the re-edited version. This will mean that the corrected version will overwrite the original, thus saving disk space and also preventing the user from mistakenly including the original version of the file.

When the corrected file has been saved, and an exit made from **REVIEW** (by pressing <aF10>), the SPSS/PC: prompt will re-appear. Try once again to **INCLUDE** the file. If it now proves acceptable, no error messages will be encountered and you will be presented with another SPSS/PC: prompt. You can now be confident that the file is free of fatal syntax errors although it may retain, for example, spelling errors in labels, or value labels which do not match the values as they were originally coded. When a **procedure command** is issued (this is usually a command requesting a statistical analysis) the relevant data will be read and an 'active file' will be created by the system.

5.7 Obtaining information about the file variables and values

Once a file has been **included** successfully you may work with it interactively by responding to the SPSS/PC+ prompt with a variety of commands. Some of the commands available are **procedure commands** and cause data to be read and analyzed. Other commands display information about the variables, and some (like **SHOW** and **SET**) are **operation commands** which provide information about the operating characteristics of the system and allow these to be changed.

Two further commands will be described. **DISPLAY** provides information about any variables which have been declared to the system. It is thus an **operation command**.

LIST displays values of the data for one or more variables. Issuing a **LIST** command therefore causes the system to read data, so **LIST** is a **procedure command** which creates an active file.

The DISPLAY command

The **DISPLAY** command produces a display of information about some or all of the variables included in the active file.

In its simplest form, the **DISPLAY** command produces a list of all the variables in the current active file, together with any variable labels that have been supplied for them. Thus:

```
SPSS/PC: DISPLAY.
```

would produce the following kind of display (and printout, if the printer has been **SET** to **on**):

```
ID        -              * No label *
SEX       -              * No label *
AGE       -              * No label *
FAC       -              Home Faculty
VAR001    -              Women equal to men
VAR002    -              New legislation
VAR003    -              Discrim. as excuse
VAR004    -              Men superior
etc.
```

If the printer was not **on** during the **DISPLAY** operation, you could repeat the operation after issuing the **SET** command 'set printer on'. If you wish to interrupt the operation of **DISPLAY** before it has run its full course, press <^C> – this is a useful general escape command and will produce an immediate interruption of many types of operation. An alternative way of obtaining a printed copy of the **DISPLAY** output is to request a direct screen print – a printed copy of the entire screen display. Search for a key on the number/function pad of the keyboard which has the label <Scr Pnt> (or something similar). Making sure that the number/function pad is switched to the function mode (there will be a key to change modes) press the <Scr Pnt> once. Provided that the printer is plugged in and has been switched on mechanically you should now obtain a printed version of the screen display. You will have to repeat this operation each time you release a new 'page' of screen display (in response to the **MORE** message).

If a variable list, or the keyword **ALL**, is included in the **DISPLAY** command, then fuller information is provided for the specified variables (or for **ALL** of the variables). Thus the following display would be produced by the command 'display sex fac.'.

```
Variable: SEX          Label:  * No label *
  Value labels follow  Type: Number  Width  1   Dec: 0   Missing 9.00
    1.00    male                                 2.00     female

Variable: FAC          Label:  Home faculty
  Value labels follow  Type: Number  Width  1   Dec: 0   Missing 9.00
    1.00    arts                                 2.00     science
    3.00    social studies                       4.00     other
```

The output for the first of these variables *sex* will be used to explain the nature of the **DISPLAY** output.

The variable name *sex* is displayed and we are informed that no label has yet been supplied for this variable. The display also suggests that *sex* is a number-type variable. This calls for a little discussion. What is meant by this is that all of the data has been coded in terms of numbers (i.e. that the data does not contain letters or other non-number characters). The fact that the variable is of the number type cannot be taken to imply that the data is truly numerical – suitable for adding, subtracting, etc. In the case of the variable *sex* we know that the numbers are in fact merely codes for 'male' and 'female' and have no valid numerical properties.

This illustrates the point that when meeting data of the number type, the SPSS/PC+ system does not distinguish between numerical and categorical variables. It is up to the user to consider the true nature of the data when requesting analyses and interpreting results. Reporting that the average value of the variable *sex* is 1.473, for example, would **not** be sensible. It cannot be emphasized too strongly that SPSS/PC+ will perform statistical analyses which are invalid and nonsensical if it is requested to do so. It is the user's responsibility to make sure that only reasonable requests are made to the system and that output is interpreted properly.

The 'Dec :' item reports how many decimal places there are to the right of the decimal point. In this case this is not a relevant issue, but the system nevertheless reports that there are 0 places to the right of the decimal point. Of more relevance is the fact that the missing value has been declared as 9. The display also reports that the value 1 has been labelled 'male', and the value 2 'female'.

The LIST command

LIST is a **procedure command** which reads the data, creating an active file, and displays the data **values** for some or all of the **variables** and for some or all of the **cases** in the file. In its simplest form (the single command word **LIST.**), the **LIST** command produces a list of values for every variable and for every cases in the active file. A **LIST** command can also specify particular variables for which the values should be displayed. Thus (using the gender attitudes file):

    SPSS/PC: LIST id sex var001 prs04.

produced the following display:

    ID SEX VAR001 PRS04

     1   1   2   2
     2   2   1   2
     3   1   5   2
     4   2   2   2
     5   2   2   1
    ...
    50   1   2   2

If a list of values is required for only **some** of the cases, the **LIST** command should specify those cases for which the data are required, using a **CASES** subcommand. The system assumes that listing of cases should begin at the first case in the file and end at the last unless it is instructed otherwise. Thus:

    SPSS/PC: LIST CASES= TO 5.

or

    SPSS/PC: LIST CASES= from 1 TO 5.

or

    SPSS/PC: LIST CASES= 5.

would all produce a list of the values of **all variables** (no variable names have been specified) for the first five cases in the file.

    SPSS/PC: LIST CASES= FROM 5.

would list the values for all cases in the file, except the first four cases.

    SPSS/PC: LIST CASES= FROM 5 TO 15.

would list the values for the 11 cases specified.

   If both a variable list **and** a **CASES** subcommand are included within a single **LIST** command they must be separated by a slash (/). Thus:

    SPSS/PC: LIST sex fac var001 to var010/ cases= from 10 to 20.

This would produce a list of values for each of the 12 variables specified for the 11 cases specified.

5.8 Using data definition files for analysis

When a data definition file has been successfully **INCLUDED** and the first procedure command is issued, the data are read in order to create an active file. Any further requested procedures then use the data from the existing active file (usually to perform a statistical analysis). One simple procedure, **FREQUENCIES**, produces tables of frequencies and percentages for the various values of a variable. A **FREQUENCIES** analysis of the data for the single variable *sex*, could be obtained by issuing the following command, for example:

```
SPSS/PC: frequencies sex.
```

This would produce a screen display and printout (if there is no printout, respond to the SPSS/PC+ prompt with the instruction **SET PRINTER ON.** and try again) of the following type:

```
SEX        Subject sex
                                              Valid     Cum.
   Value Label      Value   Frequency  Percent  Percent   Percent

Male                  1         22      44.0     44.0      44.0
Female                2         28      56.0     56.0     100.0
                              -------  -------  -------
                  TOTAL        50      100.0    100.0
```

It can be seen that the **FREQUENCIES** output (in this case from the gender attitudes survey) provides information on the value labels and the frequencies of valid (i.e. not missing) cases coded with each value. These frequencies are then expressed in terms of a percentage of the population of cases as a whole and of the population of valid cases. Valid percentages are also added sequentially to produce the cumulative percentage. As we will see in the next chapter, this represents the basic or default output from the **FREQUENCIES** procedure. Fuller output (including additional statistics, barcharts, etc.) can also be requested.

If you have a working file you may wish, at this stage, to obtain frequencies for several more variables. If you have not yet created a working file, this is a good point at which to establish a file to use while you read through the remaining chapters of this book. If you do not have data of your own at this stage, and if you have not done so already, create a full copy of the data and data definition files from the gender attitudes survey (these are reproduced in Appendix E and Appendix F). These should be saved in your own directory on the hard disk. The data definition file can then be **INCLUDED** in any future SPSS/PC+ session by issuing a command such as

```
SPSS/PC: include "gender.def".
```

When this file has been included, the contents will be available for analysis and manipulation. In the next chapter we will begin to explore in detail some of the procedures which can be used to provide statistical analyses of the dataset.

6 Statistical Analysis I:
FREQUENCIES and CROSSTABS

The two previous chapters have been concerned with file creation and editing and they have provided us with the means of creating a data definition file (and an associated data file) which can be made the active file. Data within an active file can be manipulated and analyzed by issuing the appropriate SPSS/PC+ commands interactively i.e. by keying them in response to the SPSS/PC: prompt.

Although there are many more things to learn about file creation and manipulation, this is a good point at which to introduce the use of SPSS/PC+ in performing statistical analyses.

Some readers will be very familiar with statistics, whereas others will have only a basic knowledge. For serious work it is important that you do not use SPSS/PC+ beyond your own statistical capabilities. But playing with data, particularly by using the power and convenience of SPSS/PC+, is a good way of whetting your appetite to know more about statistics. There are many good statistics textbooks, including the *SPSS^X Introductory Statistics Guide*, and the *SPSS/PC+ Manual* provides much useful information, particularly about the more complex forms of analysis.

If when using this book you meet any term with which you are unfamiliar (for example, in this chapter we use chi-square, significance, degrees of freedom, parameters, crosstabulation, variance, etc.) consult the glossary (Appendix G) for an explanation.

At this stage you will probably not be content to analyze dummy data. The analysis of a series of random numbers provides little interest. If, therefore, the data definition file you have created is of this type you might prefer to create another file, this time containing and defining real data. You can collect data in a study of your own or you can use that from the gender attitudes survey. The complete dataset obtained from this survey is provided in Appendix E, and the data definition file is provided in Appendix F. The examples given in this and many later chapters are based on this survey.

6.1 Using procedures

After creating a suitable data definition file, you can use the **INCLUDE** command:

```
SPSS/PC: include "gender.def".
```

When the SPSS/PC: prompt returns, the SPSS/PC+ system is awaiting a command. You can respond to this prompt by issuing a **procedure** command to be executed using the data from within the file. In this chapter we will consider the use of two procedure commands, **FREQUENCIES** and **CROSSTABS**.

Each procedure command runs a set part of the SPSS/PC+ program, and the procedures operate on the active file. A procedure command causes the data to be read from the active file and, usually, subjects the data to some type of statistical analysis. Most procedure commands can be supplemented with one or more subcommands which allow users to tailor the procedure to their own needs. Two of the most useful, and generally used, subcommands are **OPTIONS** and **STATISTICS**. The **OPTIONS** subcommand is frequently used to specify how missing data should be dealt with and to request a particular form for the output display or printout. The **STATISTICS** subcommand allows the user to request statistical information additional to that provided by the procedure by default.

We will start to use procedure commands (including subcommands) interactively with the SPSS/PC+ system (i.e. in response to the SPSS/PC: prompt). Later (in Chapter 9) we will see that procedure commands and subcommands can also be written into a separate file or added directly into data definition files (such files are referred to as command files). At this stage, however, we will assume that the procedure commands and subcommands are to be issued interactively.

After issuing a complete command, ending with a terminator period (.), a number will appear in the top right hand corner and will change rapidly, indicating that cases are being processed and analyzed. When this number reaches the total number of cases in the file, and after a short pause during which further processing is taking place, the first page of results will be displayed on the screen. If the printer has been set to on then the first page of results will also be printed.

After the first page has been displayed, the operation will be halted. The message **MORE** will be presented at the top right hand corner of the screen and a high-pitched beep will sound. Pressing any key on the keyboard will release the next page of information.

If the user wishes to examine the results of the procedure in the form of printed output, rather than as successive screen displays, then the pause between pages (accompanied by the **MORE** message) will be redundant.

The pause facility can be switched off (so that all of the output is displayed and printed without a pause) by using the **SET** command (*before* issuing the procedure command) to 'set MORE off.':

```
SPSS/PC: set more off.
```

We will now consider the use of two SPSS/PC+ procedures for statistical analysis, **FREQUENCIES** and **CROSSTABS**.

6.2 The FREQUENCIES procedure

It is best to begin the analysis of a dataset by examining the basic features of single variables. For coded variables we might wish to know the frequency distribution over the range of different values; this will allow us to examine the composition of the sample in terms of sex, age and other such parameters and to find out how many cases have any particular value for a variable. We might also require basic statistical information such as the mean and standard deviation of individual variables. The **FREQUENCIES** procedure produces tables of frequency counts for the values of any variable(s) specified and can also be used to display a variety of statistics for such variables.

To use **FREQUENCIES** we first need to make the relevant file active, and we do this by issuing an **INCLUDE** command, e.g.

```
SPSS/PC: include "gender.def".
```

The SPSS/PC+ prompt will again be presented and we can respond with the procedure command:

```
SPSS/PC: frequencies sex.
```

When <RETURN> is pressed there will be a pause while SPSS/PC+ processes the data. The following results table will then be displayed:

```
SEX        Subject sex
```

Value Label		Value	Frequency	Percent	Valid Percent	Cum Percent
Male		1	22	44.0	44.0	44.0
Female		2	28	56.0	56.0	100.0
		TOTAL	50	100.0	100.0	

```
Valid Cases    50    Missing Cases    0
```

We can see that the total population of valid cases is 50 and that 22 of these are male and 28 are female. The table also includes the frequency of each *sex* group as a percentage of the total population (valid + missing) and of the total valid (only) population. The cumulative percentage is also displayed.

Several variables can be listed within a single **FREQUENCIES** command (and the **TO** keyword can be used). Thus the following list produces a frequency table for each of the eight variables listed:

```
SPSS/PC: frequencies sex  var001 to var006 st02.
```

If such a command is issued without the terminator period, and <RETURN> pressed, there will be a pause while SPSS/PC+ processes data and a continuation prompt (:) will then be displayed. This is an opportunity to enter a subcommand before the output is obtained.

Statistical information can be obtained by issuing a **STATISTICS** subcommand. This specifies which statistics are to be displayed by quoting one or more keywords (for example **stddev** to obtain the standard deviation).

The **STATISTICS** subcommand is specified after the main command following a slash (/) and takes the form:

```
: / statistics = median stddev minimum.
```

If this is the final subcommand to be issued then it should end with the terminator period (as above).

Thus the complete command would be entered in two stages, e.g.

```
SPSS/PC: frequencies sex  var001 to var006 st02  <RETURN>
```

and then:

```
: / statistics = median stddev minimum.
```

There are many ways of finding out which **STATISTICS** are available for any procedure. Lists of the available **STATISTICS** are provided in the *SPSS/PC+ Manual* and on the *SPSS/PC+ Reference Card* which is included as an addition to the *SPSS/PC+ Manual*. A selected list of the more frequently used **STATISTICS** (and **OPTIONS**) is also provided at the end of this book (Appendix J). In addition to these off-line information sources, on-line help can be obtained by using the SPSS/PC+ **HELP** facility in response to the SPSS/PC: prompt. Thus:

```
SPSS/PC: help frequencies statistics.
```

will provide a screen display of the **STATISTICS** available for the **FRE-QUENCIES** procedure, including a keyword which may be used to request an available statistic.

The **STATISTICS** available with **FREQUENCIES** include measures of central tendency (mean, mode and median), measures of dispersion (standard deviation, variance), and measures of the shape of the frequency distribution (skewness and kurtosis). For a general explanation of such statistical terms see the glossary, Appendix G. The **FREQUENCIES STATISTICS** are also briefly described in Table 6.1. The keyword item in the table refers to the word used to request the statistic within the **STATIS-TICS** subcommand.

In specifying which statistics are to be performed on the data, the keywords **ALL** and **NONE** can be used. If a **STATISTICS** subcommand is issued without specifying any particular statistics then certain default statistics – MEAN, STDDEV, MINIMUM and MAXIMUM – are displayed. If no **STATISTICS** subcommand is included then no statistics are displayed.

It should be noted that data relating to particular variables will be suitable only for *some* of the analyses available, but that SPSS/PC+ will, at the user's request, perform many totally invalid analyses on data. Thus if a specific request is made, or a relevant default not over-ridden, SPSS/PC+ will happily produce a mean (average) for a categorical variable. Thus, using the data from the gender attitudes survey, a mean of 1.56 was obtained for the variable *sex*. If we had chosen instead to code males as 2 and females as 1 – an equally arbitrary strategy – the mean for the same population would then have been represented as 1.44).

Such figures, however, are clearly quite meaningless. In requesting a mean (or standard deviation, or median, etc.) for such data we have completely ignored the vital difference between *coded* variables (such as *sex*) and genuine *numeric* variables (such as *salary*). The point is that SPSS/PC+ is not aware of the type of data in use. It will perform the analyses it is asked to perform, as long as certain basic criteria are fulfilled. It has no appreciation, in cases such as the example we have given, of whether data are numeric or categorical. The responsibility for the appropriateness of the analyses performed, therefore, rests with the user rather than with the system.

Although there are many clear-cut instances in which particular types of analysis are appropriate or inappropriate, there are situations where opinions on the appropriateness of employing particular procedures differ sharply. Some statisticians, for example, are happy to use a full range of statistical tests on seven-point scale data whereas others consider such data to be categorical rather than numerical and believe that arithmetic operations should not be performed.

There is no **OPTIONS** subcommand for the **FREQUENCIES** procedure, but there are various ways of tailoring the output. Thus barcharts and histograms can be requested. For example, to request a simple barchart:

```
SPSS/PC: frequencies fac sex var001 to var006 prs05
      : / barchart.
```

This would produce, as well as the normal display information, a barchart

Table 6.1 Frequencies statistics

Keyword	Name	Description
MEAN	Mean	The arithmetic average
STDDEV	Standard deviation	A measure of the spread of the frequency distribution
MODE	Mode	The most frequent value for this variable within the data
MINIMUM	Minimum	The lowest value
MAXIMUM	Maximum	The highest value
RANGE	Range	The difference between the lowest value and the highest value
SEMEAN	The standard error of the mean	A measure of the confidence we can have that the average based on this sample is a true representation of the average of a larger theoretical population
VARIANCE	Variance	Another measure – along with STDDEV – of the spread, or variation, within the data
SKEWNESS	Skewness	A measure of how lop-sided the data are – i.e. a tendency for more data to occur towards the lower or upper end of the range rather than being evenly distributed or balanced equally on either side of the mid-point of the range
SESKEW	The standard error of the skewness	A measure of the confidence we can have that the skewness of this sample is a true representation of the skewness of a larger theoretical population
MEDIAN	Median	The value which is a mid-point, dividing the population into 50% below the median and 50% above the median
KURTOSIS	Kurtosis	The extent to which the data frequencies peak around the middle value or are flat across the range of values
SEKURT	The standard error of the kurtosis	A measure of the confidence we can have that the kurtosis of this sample is a true representation of the kurtosis of a larger theoretical population
SUM	Sum	The sum of all the data points for the variable

for each of the variables listed on the **FREQUENCIES** command. The raw frequency of each value for that variable is plotted on a horizontal axis, and value labels are included in the output. The output from the **FREQUEN-CIES** procedure (including that obtained from the **BARCHART** and **HIS-TOGRAM** subcommands) can be tailored in a number of ways. The **SET** command, for example, can be used, *before* the procedure command is issued, to change the symbol used to print a histogram or barchart. Full information is available in the *SPSS/PC+ Manual* (pp. C65–C69; C172–C183) and on-line guidance can be obtained by using the SPSS/PC+ **HELP** facility, e.g.

```
SPSS/PC: help frequencies barchart.
```

If two or more subcommands are used within the same command they must be separated by a slash, e.g.

```
SPSS/PC: frequencies fac sex var001 to var006 prs05
       : / statistics = mode range
       : / barchart.
```

Only the final subcommand is followed by the terminator.

6.3 The CROSSTABS procedure

We often wish to know the frequencies of values with respect to two or more coded variables simultaneously. Thus if we have information on the sex of our respondents and on their marital status we might wish to display a table which provides the frequencies of married males, married females, divorced males and divorced females. This would enable us to examine the question of whether within our sample the proportion of males who are divorced is greater than the proportion of females who are divorced. If this proved to be the case then we might conclude that (for our sample) sex and the state of being divorced are not independent, or that there is an association between sex and divorced status. A crosstabulation, or joint frequencies table, which will display such information in a convenient form can be obtained by using the **CROSSTABS** procedure. In its simplest form it simply names a variable, uses the keyword **BY** and then names a second variable. Thus:

```
SPSS/PC: crosstabs fac BY sex.
```

A crosstabulation table displays the joint distribution of two (or more) variables each with a limited number of (usually categorical) values. In the above example, because the variable *sex* can take either of two values, and *fac* can take one of three values (in the gender attitudes survey arts, science and social studies), a crosstabulation of the two variables will be a 2×3 table with six cells:

```
                      Male          Female
                 ........................... Row Total
                 :         :         :
       Arts      :    6    :    12   :    18
                 :.........:.........:
                 :         :         :
     Science     :    8    :    4    :    12
                 :.........:.........:
                 :         :         :
  Soc. Studies   :    8    :    12   :    20
                 :.........:.........:

       Column        22            28      TOTAL   50
       Total
```

The three rows correspond to the three valid values for *fac* and the two columns correspond to the valid values for *sex*). There are thus six cells. Reading the information from the table we can see that there are 50 subjects, 22 male and 28 female. Of our population 18 are from the arts faculty, 12 are from the science faculty and 20 are from the social studies faculty. From the information within the cells we can see, for example, that there are 12 female arts students in the population, and 8 male science students.

The above table was obtained by using the **CROSSTABS** command :

```
SPSS/PC: crosstabs fac BY sex.
```

Had we specified the alternative order (*sex* by *fac*) then *sex* would have been displayed as rows and *fac* as columns:

```
SPSS/PC: crosstabs sex BY fac.

  FAC              Arts      Science   Soc. studies
                 .................................... Row Total
  SEX            :        :        :        :
       Males     :    6   :    8   :    8   :    22
                 :........:........:........:
                 :        :        :        :
     Females     :    12  :    4   :    12  :    28
                 :........:........:........:

     Column          18        12       20      TOTAL 50
     Total
```

The basic form of a **CROSSTABS** command is

   CROSSTABS variable name (or list) **BY** variable name (or list), e.g.

```
SPSS/PC: crosstabs prs01 to prs10 BY sex fac.
```

This will produce 20 separate crosstabulations (i.e. each of the 10 variables *prs01* to *prs10* will be crosstabulated with *sex* and, separately, with *fac*).

So far we have considered only crosstabulations employing two variables. It is also possible to request crosstabulations which display the frequency distributions according to three or more variables. Thus :

```
SPSS/PC: crosstabs fac BY sex BY st04.
```

would give a single three-dimensional table, in which each cell would include the number of cases for one faculty, one sex and one value of the variable *st04*. Thus with three faculties, the two sexes, and two possible (valid) values for *st04*, there would be:

$3 \times 2 \times 2 = 12$ cells.

Up to 10 dimensions can be specified in any one command. However, as the number of cells increases the average number of cases per cell will obviously decrease and a low number within one or more cells (and especially, the presence of empty cells) will prohibit the legitimate use of many of the statistical tests available for **CROSSTABS**.

If no **OPTIONS** or **STATISTICS** subcommand is included in the overall procedure command, then the procedure will operate with all of its defaults in effect. By default, the **CROSSTABS** output includes tables with frequency counts (but not percentages) in each cell, marginal counts and percentages, the values, variable labels and value labels, and the numbers of valid and missing cases.

Thus the following table presents the complete default output from the command:

```
SPSS/PC: crosstabs sex BY prs10.
```

```
Crosstabulation:       SEX        Subject sex
                    By PRS10      Trust difficult (SELF)

            Count  |TRUE    |FALSE   |
  PRS10->          |        |        |    Row
                   |      1 |      2 |  Total
  SEX      --------+--------+--------+
              1  |      8 |     14 |     22
  Male           |        |        |   44.0
                 +--------+--------+
              2  |     12 |     16 |     28
  Female         |        |        |   56.0
                 +--------+--------+
          Column       20       30       50
           Total     40.0     60.0    100.0

Number of Missing Observations =         0
```

The **CROSSTABS** command may include both **OPTIONS** and **STATISTICS** subcommands. SPSS/PC+ provides no fewer than 19 different

OPTIONS for the **CROSSTABS** procedure, and the use of one or more of these will affect the information included within the crosstabulation display. Up to 11 additional statistical descriptors (or statistical tests) can also be requested on a **STATISTICS** subcommand by citing appropriate reference numbers.

Contrast the display in the table above with the equivalent display obtained when **OPTIONS 3, 4, 5** and **14** are requested. The command:

```
SPSS/PC: crosstabs sex BY prs10
       : /options = 3 4 5 14.
```

produces the following output :

```
Crosstabulation:       SEX        Subject sex
                    By PRS10       Trust difficult (SELF)

              Count  I
              Exp Val I
              Row Pct ITRUE     IFALSE    I
  PRS10->     Col Pct I         I         I    Row
              Tot Pct I      1  I      2  I  Total
  SEX         --------+--------+--------+
                  1   I      8  I     14  I     22
  Male              I    8.8  I   13.2  I   44.0
                    I   36.4  I   63.6  I
                    I   40.0  I   46.7  I
                    I   16.0  I   28.0  I
                    +--------+--------+
                  2   I     12  I     16  I     28
  Female            I   11.2  I   16.8  I   56.0
                    I   42.9  I   57.1  I
                    I   60.0  I   53.3  I
                    I   24.0  I   32.0  I
                    +--------+--------+
              Column       20       30       50
              Total      40.0     60.0    100.0
```

Each cell now contains five items of information and a guide to the various items is provided is the top left corner. The first item in each cell is the cell count (as in the default table). The second item is the frequency which would be expected in that cell if the two variables (in this case *sex* and *prs10*) were statistically independent. This item has been requested by **OPTION 14**. The remaining items are percentages – the cell frequency is expressed as a percentage of the row total, the column total and the population total. These items have been requested by **OPTIONS 3, 4** and **5**.

You will recall that there are many sources of information on the **STATISTICS** and **OPTIONS** available for a procedure; remember, particularly, that the **HELP** command, e.g. help crosstabs options, can be also used to obtain such information.

When requesting additional **STATISTICS** the user may also use the keyword **ALL**. However, **ALL** cannot be used within the **OPTIONS** sub-command because many procedures allow, as alternatives, **OPTIONS** which are mutually incompatible.

Thus a complete **CROSSTABS** procedure command, with these two types of subcommands included, would take the form:

```
SPSS/PC: crosstabs sex fac BY st01 to st08 / sex BY yr
       : /OPTIONS= 3  4
       : /Statistics= 1 3 9 10.
```

This will yield 17 separate crosstabulation tables (notice too in this example how either upper or lower case characters may be used).

6.4 Inferential statistics: the use of chi-square

Some statistics are merely summaries or descriptions of the data. They may cite an average, for example, or summarize the spread of the data. Such statistics (which include the mean, median, mode, range, standard deviation, kurtosis, etc.) are known as *descriptive statistics*. Other statistics allow us to draw conclusions about whether, for example, two samples differ, or whether there is an association between two variables which cannot simply be attributed to chance. Such statistics emerge as the result of a statistical test, and they allow us to make an inference. Hence they are called *inferential statistics*. Chi-square is an example of an inferential statistic and we will meet with many others.

When the value for such a statistic has been calculated we usually wish to know whether that value represents something significant. It is in the nature of such statistics that it is possible to compute the probability that any value would emerge from essentially random data by chance. Generally, the higher the value of such a statistic the less frequently it would emerge by chance. Thus if a high value (of chi-square, for example) is obtained using the data entered into the statistical equation then it is likely that this reflects some real pattern in the dataset rather than mere chance.

A common criterion for deciding whether such there is a pattern in the data involves applying a 1 in 20 rule. If the value obtained for the statistic is higher than we would expect to emerge by chance alone on 5 per cent (1 in 20) or fewer occasions then we infer that our value is not the result of chance but is likely to reflect a real effect (or pattern). We say that such a result is *significant*. We might wish to play safer and adopt a more stringent criterion by using a 1 in 100 rule. This would demand that a higher value of the statistic be obtained (such as would occur by chance on only 1 per cent of occasions) before inferring that a pattern or effect has been demonstrated. Typically, when inferential statistics have been calculated, the SPSS/PC+ output also provides a value for the probability of that value. Thus if we are using a 5 per cent probability criterion (which could be written 0.05) the probability values 0.035 and 0.021 (i.e. even less probable than 0.05) would be sufficient for us to conclude that the pattern or effect was significant. Probability values of 0.06 or 0.62, however, would not be sufficiently low (the observed pattern would not be sufficiently improbable) for us to conclude that an effect was present. We would not be able to conclude that there was no effect but would merely have to admit that our data did not show a significant effect.

If the 1 per cent (0.01) criterion is adopted then SPSS/PC+ would have to return a probability value lower than 0.01 (e.g. 0.003 or 0.0019) before we accepted that a significant effect had been demonstrated.

The **STATISTICS** available for the **CROSSTABS** procedure include a number which assess the degree of association between the variables specified by the command. Thus for chi-square (**STATISTIC 1**), a significant result indicates that there is a relationship between the row variable and the column variable. Thus if, in the *sex* by *fac* analysis, a chi-square test had been requested and a significant result obtained, this would have indicated that, within our data, there was a significant association between respondents' sex and their home faculty. An examination of the crosstabulation might have revealed, for example, that more females tended to be from the arts and social studies faculties and more males from the science faculty, and we would know (because of the statistical significance of the chi-square) that the observed faculty distribution of the sexes was unlikely to have been a result of mere chance.

An analysis of the following table **fails** to indicate a **significant** association between *sex* and *prs07* ('trouble sleeping'):

```
     SEX     Subject sex
  By PRS07   Trouble sleeping (SELF)

                        TRUE       FALSE      Row
                   .....................     Total
                    :        :         :
         Male      :   7     :   15    :      22
                   :........:.........:
                    :        :         :
         Female    :   6     :   22    :      28
                   :........:.........:

         Column        13        37
         Total                        TOTAL 50

Chi-square   D.F.   Significance   Min E.F.   Cells with E.F. < 5

   .25667      1        .6124         5.720       None
   .69120      1        .4058      (Before Yates Correction)
```

This output table needs some explanation. We see that two values have been calculated for chi-square. This is because the table is a 2×2 cross-tabulation. The bottom line includes a normal chi-square calculation (i.e. before Yates correction). The penultimate line includes an alternative calculation using a modified formula which, for technical mathematical reasons, incorporates Yates Correction. When the correction factor has been applied (and it is applied only to 2×2 tables) we can see that the effect is to reduce the value of chi-square and hence to increase the probability estimate (to 0.6124). Increasing the probability in this way makes it less likely that a significant result will be obtained, and the use of Yates correction therefore provides a more conservative (or safer) value for chi-square. There is some controversy about whether it is necessary to apply this correction or not, but some critics do insist that it should be employed when a chi-square analysis of 2×2 tables is performed.

In addition to the two chi-square values, with their associated probabilities, the table includes values for D.F. (degrees of freedom – see glossary). It also reports on an interim calculation of the expected frequencies of the cells in the table. The chi-square formula calculates expected frequencies and then compares these with the actual (observed) frequencies in the cells. If the differences between the expected and observed values are high then a high value is obtained for chi-square. The information in the table includes the minimum expected frequency and the number of cells which have an expected frequency of less than 5. If there are cells with a lower expected frequency then some statisticians would doubt the validity of the chi-square statistic obtained.

The chi-square value for the above table (after applying Yates correction) is 0.257. This corresponds to a probability of 0.6124 which, being far higher than the 0.0500 threshold for significance, indicates that – according to our dataset – one sex is not more likely to have sleeping difficulties than the other. This reflects the fact, observable within the crosstabulation, that roughly the same *proportion* of males as females say that they have trouble sleeping.

In the next table, however, we see that there is an apparent association between the two variables.

The information in this table indicates that a higher proportion of males than females in the sample claimed that they thought about sex 'often'. The chi-square value for the table (with Yates correction applied) is 3.977 and the probability associated with this value is 0.0461. Thus the probability of such a set of data emerging purely by chance is just less than the 0.05 (or 5%) criterion and we can therefore conclude that the data shows a significant relationship between the subject's sex and the likelihood of their claiming to think about sex often.

```
      SEX      Subject sex
   By PRS02    Sex thoughts often (SELF)

                         TRUE        FALSE      Row
                      ....................     Total
                      :        :        :
           Male       :   15   :    7   :   22
                      :........:........:
                      :        :        :
           Female     :   10   :   18   :   28
                      :........:........:

              Column       25          25
              Total                        TOTAL 50
```

Chi-square	D.F.	Significance	Min E.F.	Cells with E.F. < 5
3.97727	1	.0461	11.000	None
5.19481	1	.0227	(Before Yates Correction)	

Other **STATISTICS** available with **CROSSTABS** also measure the degree of association between the crosstabulated variables. They make different assumptions about the data, however, and for some data some of the tests are considered more suitable than others. The choice of test will depend on the needs of the user and on certain conditions within the data.

Thus the legitimate use of the chi-square test demands that no cell is empty and the use of chi-square on a table containing any cell with an expected frequency of less than 5 is controversial. Once again it is worth making the point that SPSS/PC+ will often calculate invalid statistics. Thus it will perform chi-square tests even if certain assumptions of the test are infringed.

Although SPSS/PC+ provides a wide range of tests, there would never be a situation in a real research study in which all of the available statistics were appropriate or necessary. The needs of most users will probably be satisfied by employing only one or two of the statistics available for **CROSSTABS**, and chi-square will almost certainly be found to be one of the more useful.

In the next chapter we consider the use of two further statistical procedures, **MEANS** and **T-TEST**.

7 Statistical Analysis II: MEANS and T-TEST

In this chapter we continue our examination of SPSS/PC+ procedures by introducing two more procedure commands, **MEANS** and **T-TEST**. Remember to consult the glossary (Appendix G) to find the meaning of any terms you do not understand.

7.1 The MEANS procedure

The **MEANS** procedure groups cases by their value on one or more independent variables (for example, *sex*) and then displays statistics for numerical dependent variables (for example *var001*) for each of the groups which has been formed (say, 'male' and 'female'). Thus the case population is said to be broken down into groups (and possibly subgroups). The procedure produces means (averages), standard deviations and a count of the number of valid cases in each of the groups. Thus, the command

```
SPSS/PC: MEANS var001 to var005 BY prs04.
```

would produce five tables, one for each of the variables specified to the left of the **BY** keyword. In each table the means, standard deviations and counts of each of the two 'groups' (one group for each of the two values of *prs04*) would be given, and the same statistics would also be given for the population as a whole. The dependent variables (those to the left of the **BY** keyword) must be numerical (rather than categorical) if the averages and standard deviations are to be at all meaningful. The independent variables (those to the right of the **BY** keyword), on the other hand, will usually be categorical in nature (like *sex* or *prs04*). From the command

```
SPSS/PC: MEANS var001 to var010 BY sex.
```

we would obtain 10 tables of the means, standard deviations and valid counts for variables *var001* to *var010* for males and females separately. The above command produced the following output for *var008*:

```
Summaries of    VAR008    Men more sexy
By levels of    SEX       Subject sex

Variable        Value  Label        Mean      Std Dev   Cases

For Entire Population                4.7600    1.8905    50

SEX             1    Male           3.8636    1.8334    22
SEX             2    Female         5.4643    1.6439    28

Total Cases  =  50
```

It is possible to break down the population by more than one variable (i.e. to use more than one *independent* variable). Thus the command:

```
SPSS/PC: MEANS var001 to var008 BY sex BY prs01.
```

would produce eight separate tables, one for each of the dependent variables *var001* to *var008*. Each table would include statistics for the population as a whole, for the male group and the female group separately, and for each of the two subgroups (*prs01* value 1, true, and *prs01* value 2, false) **within each sex group** (i.e. four subgroups altogether). Thus each possible combination of values of the independent variables defines a group or subgroup. The above command produced the following as one of the eight output tables:

```
Summaries of    VAR007    Women more sensitive
By levels of    SEX       Subject sex
                PRS01     Depressed often (SELF)

Variable        Value  Label        Mean      Std Dev   Cases

For Entire Population                4.0400    1.8290    50

SEX             1    Male           4.0000    1.9518    22
  PRS01         1    TRUE           4.4000    1.9551    10
  PRS01         2    FALSE          3.6667    1.9695    12

SEX             2    Female         4.0714    1.7623    28
  PRS01         1    TRUE           3.6842    1.9164    19
  PRS01         2    FALSE          4.8889    1.0541    9

Total Cases  =     50
```

An examination of this table suggests (although further statistics would be needed to support this interpretation) that while there is no overall difference in the extent to which males and females believe that women are more sensitive than men, males who get depressed are less likely to believe this ('women are more sensitive') than males who do not get depressed. On the other hand, females who get depressed seem somewhat *more* likely to believe that females are sensitive than are females who do *not* get depressed.

Up to five **BY** words can be used in any command, and therefore the number of subgroups which can be formed using a **MEANS** command may be very large. As more subgroups are specified, however, the average number of cases which fall into each subgroup will diminish. Eventually the numbers in particular groups might become so small that the standard deviations, in particular, will become meaningless.

The order in which independent variables are named affects the nature of the display of the table (but not the values contained within it). The first-named independent variable is presented as the major classification criterion, with later-named independent variables presented as increasingly minor.

Two or more lists of variables can be included on separate subcommands within a single **MEANS** command. For example:

```
SPSS/PC: MEANS var001 to var004 BY prs04 BY sex
       : /var007 var009 var010 by st06.
```

This example would produce seven separate tables (four requested by the first subcommand and three by the second).

The command:

```
SPSS/PC: means var001 to var003 BY prs01 prs02 prs03.
```

will produce nine separate tables: *var001* with *prs01*, *var001* with *prs02*, *var001* with *prs03*, *var002* with *prs01*, etc.

In the default case, means, standard deviations and counts are included for the entire population, for each group and for each subgroup. The output is fully labelled with variable labels and value labels. If a case has a missing value for one or more of the variables involved in a table specification then the case is omitted from all statistics calculated for the table.

OPTIONS: Information regarding **OPTIONS** may be obtained by the command:

```
SPSS/PC: help means options.
```

Eleven **OPTIONS** are available for the **MEANS** procedure. They relate to the use of missing data, the suppression of names and labels, and the display of group sums and group variances. For example, **OPTION 1** requests that missing cases be included in the calculations, and **OPTION 3** causes all labels to be suppressed.

STATISTICS: Two additional **STATISTICS** are available for **MEANS**:

STATISTIC 1 requests that a one-way analysis of variance (see Chapter 12) be performed using the first-named independent variable.

STATISTIC 2 requests a test of linearity. This again involves a one-way analysis of variance using the first-named independent variable, and linearity is assessed.

A full **MEANS** command might take the form:

```
SPSS/PC: means var002 to var006 BY sex BY prs06
       : /var002 var004 var007 BY st03
       : /options= 3
       : /statistics= 1.
```

This requests that eight tables (5 + 3) be displayed without labels (**OPTION 3**) and that a one-way analysis of variance be performed on the data within each table.

7.2 The T-TEST procedure

The **T-TEST** procedure is used to compare the means (average values) of two variables for a single group of cases, or of the same variable for two different groups. It should be used with numeric data and *not* with coded variables. There are two forms of the *t*-test.

(a) **An independent samples t-test** is used to compare two groups on a single specified variable. Such a test can reveal, for example, whether our male subjects have a significantly different *age* score than our female subjects.

Although a coded variable, *sex*, is involved in this comparison, the coded values are used here only to define the two groups. Apart from using the codes to form the groups, the system makes no use of the values 1 for 'male' and 2 for 'female' in the statistical analysis. The computation of the *t*-test statistic involves only the numbers that represent *age* in years.

(b) **A paired samples t-test** (sometimes referred to as a paired *t*-test, or a correlated *t*-test) is used to assess whether, for a single group of subjects, the score on one variable is significantly different from the score on another variable. Thus it could be used to determine whether respondents' values for *var001* (the 1–7 agreement rating supplied in response to the statement '**Women are equal to men in every way**') are significantly different from the same respondents' values for the variable *var010* – reflecting their level of agreement with the statement '**Most men feel very threatened by highly intelligent women**'. It should be noted that some statisticians object to the use of seven-point scale data in such analyses although others are satisfied that such data need not infringe the assumptions of the *t*-test.

When a *t*-value has been calculated we need to know the associated probability. Low *t*-values may reflect the effects of mere chance. High *t*-values are more likely to reflect a real difference between the groups (or between the variables).

The probability value associated with any particular *t*-value will depend on how many 'degrees of freedom' are involved and also on the nature of the hypothesis being tested. If the hypothesis is directional (e.g. males are older than females) then a lower value will be needed to obtain a significant probability value than if the hypothesis is of a non-directional kind (e.g. there is a difference, in one direction or the other, between the ages of males and females). A directional hypothesis is tested with a one-tail test and a non-directional hypothesis with a two-tail test. SPSS/PC+ output includes two-tail probabilities. The equivalent one-tail probability is found simply by dividing the two-tail probability by 2.

The independent t-test

The SPSS/PC+ procedure command for an **independent** *t*-test starts by specifying how the two comparison groups are to be formed. Thus if we wish to form two separate groups, one consisting of the male respondents and the other consisting of the female respondents, we must specify that the groups variable is *sex* and then instruct the system how to form groups using that variable.

Thus to request an independent *t*-test on *age*, for the two sexes, we would use the command:

```
SPSS/PC: t-test groups=sex (1,2) / variables = age.
```

The first part of this command (before the slash, /) instructs the program to define two groups on the basis of the variable *sex*. Cases with the value 1 on the variable *sex* (in this case males) will form one group and those with the value 2 (in this case females) will form the other group. If only one value is specified within the brackets then cases with this value form one group and all other cases are treated together as a second group.

The second part of the command (after the slash, /) tells the program to compare the two defined groups on the variable(s) listed (in this case the single variable *age*).

Here is another example of an independent *t*-test command:

```
SPSS/PC: t-test groups= prs06 (1,2)
         :/ variables = age var006 to var008.
```

This would lead to the formation of two groups, one consisting of those who said (*prs06* value 1) that they enjoyed taking risks, and one consisting of those who said that they didn't (*prs06* value 2). These two groups would then be compared on the variable *age*, and on their responses to three of the attitude questions (*var006* to *var008*).

The above command revealed that the group of respondents who said that they enjoyed taking risks differed in their degree of agreement with the statement '**Women are emotionally more sensitive than men**' (*var007*) than those who said that they did *not* enjoy taking risks (*t*-value = 2.62, two-tailed probability = 0.012).

The output which revealed this finding took the following form:

```
Independent samples of   PRS06      Enjoy risks (SELF)

Group 1:  PRS06  EQ 1              Group 2:  PRS06  EQ 2

t-test for:  VAR007    Women more sensitive
```

	Number of Cases	Mean	Standard Deviation	Standard Error
Group 1	28	4.6071	1.707	.323
Group 2	22	3.3182	1.756	.374

```
                  : Pooled Variance Estimate :   Separate
                  :                           :   Variance
  F     2-Tail    :    t    Degrees of 2-Tail :   Estimate
Value   Prob.     :  Value   Freedom   Prob.  :
                  :                           :   (see below)
 1.06    .878     :  2.62       48      .012  :
```

The default output from the independent *t*-test procedure (as above) displays the means, standard deviations, standard errors and counts of valid cases for each group. In addition, the degrees of freedom are given and there are two *t*-values, one based on the 'pooled variance estimate' and one based on the 'separate variance estimate'. This is explained below. Each of the *t*-values is then expressed in terms of its probability – the two-tailed probability of the difference between the two group means. The one-tailed probability is simply half of this value; for the above example it would be 0.006. Cases which have missing values for the dependent variable (in this case *var007*) are *not* included in the calculation. The display also includes the variable labels.

A variance ratio (**F-test**) value is given, along with its probability value. This test compares the variances of the two samples to determine whether these variances are significantly different. The *F*-test result is used in the following way. If the *F*-value is significant (i.e. associated with a probability of *less than* 0.05) then the *t*-value calculated from the **separate variance estimate** should normally be used (the relevant information, which *would* be included in normal SPSS/PC+ output is not included in the above table). If, however, the *F*-value is *not* significant then the *t*-value calculated from the **pooled variance estimate** should normally be used. (In most cases the two *t*-values are in fact very similar.) Thus in the above example the *F*-value (1.06) is not significant (probability = 0.878) and we can therefore conclude that the two variances are *not* significantly different. Thus the calculation based on the pooled variance estimate is used to draw conclusions about the means of the two groups. In this case we take 2.62 to be the relevant value of *t* and we base our conclusions on that.

Those who are relatively new to statistics may find some of this discussion difficult. A fuller (and gentler) explanation of the independent *t*-test will be found in most introductory statistics textbooks.

The **OPTIONS** available for the independent *t*-test are few.

OPTION 1 specifies that missing values should be *included* in the analysis (they are normally excluded, of course).

OPTION 2 specifies that a case (subject) should *not* be included in *any* of the *t*-test analyses specified by the command if that case has missing data for *any* of the variables specified on the *t*-test command. Thus with the following instruction:

```
SPSS/PC: t-test groups= sex (1,2)
       : / variables= var001 to var006
       : / options= 2.
```

any case with a missing value for any one or more of the listed variables would be excluded from *all* of the six analyses requested by the command.

The effect of **OPTION 2** can be illustrated by running the above command on the gender attitudes survey data, once with the **OPTION 2** subcommand and once without it. Each output display (there will be 12 in all) will include a report of how many male and female cases have been processed for the particular analysis. With **OPTION 2** requested the numbers can be seen to remain constant throughout the series of six displays, each analysis being based on 21 male and 26 female cases. These are the cases within the full dataset (of 22 males and 28 females) for which complete valid data are available for the variables *var001* to *var006*. When no **OPTIONS** subcommand is issued, however, the numbers of cases analyzed, as reported in the series of six displays, vary between 21 and 22 for the males and 27 and 28 for the females. These analyses are based on all cases which have a valid value for the particular variable being analyzed.

This latter strategy of deleting only those cases for which data is missing for a variable directly involved in a specific analysis is known as pairwise deletion. This is generally the default strategy used by SPSS/PC+ where there are cases with missing data values. The alternative strategy which follows from specifying **OPTION 2** within a **T-TEST** command is known as listwise deletion.

OPTION 3 specifies that variable labels should *not* be included on the output.

No additional statistics are available for the **T-TEST** procedure.

There are certain limitations with the independent *t*-test which should be borne in mind if complex or large-scale analyses are to be performed:

- Only *one* **GROUPS=** statement can appear in a single **T-TEST** procedure command, and the **GROUPS=** statement can name only one variable (however, a sequence of several commands can of course be submitted to the system).
- No more than 50 variables can be specified on single **T-TEST** procedure command (and they should all, of course, be numeric).

- Only one each of the **GROUPS**, **VARIABLES** and **OPTIONS** subcommands can be included within a single independent **T-TEST** command.

The paired t-test

The **paired samples** *t*-test can tell us whether, for a single group of subjects, the score on one numeric variable is significantly different from the score on another numeric variable.

The command for a **paired** *t*-test takes the form:

```
T-TEST PAIRS= var001 WITH var002.
```

Thus a paired *t*-test could be used to assess whether the degree of agreement respondents have expressed towards one statement (as represented, for example, by *var001*) is significantly different from that expressed towards another (for example, that represented by *var010*).

The **T-TEST** command used to specify such a comparison would be:

```
SPSS/PC: t-test pairs= var001 with var010.
```

The keyword **WITH** can be omitted if you wish to perform paired *t*-tests between every pair in a variable list. Thus:

```
SPSS/PC: t-test pairs= var001 var010.
```

is equivalent to the command immediately above. The command:

```
SPSS/PC: t-test pairs= var001 to var006.
```

will produce 15 paired *t*-test comparisons (each variable, *var001* to *var006*, will be paired with every other variable in the group).

Lists can be included in the command on one or both sides of **WITH**:

```
SPSS/PC: t-test pairs= var001 to var004 with var006 to var010.
```

Here each of the four variables in the first part of the list will be paired with the variables *var006* to *var010* (making 20 paired *t*-tests in all). Note that in using this command the differences between pairs of the first four variables will *not* be calculated.

Multiple **PAIRS** statements can occur within a single **T-TEST** command. Separate statements are separated with a slash (/) and the **PAIRS** keyword need not be included after the first time. Thus:

```
SPSS/PC: t-test pairs= var001 to var006 with var008
       : /var001 with var002
       : /var003 with var005 var007 to var010.
```

This would yield 12 paired *t*-test analyses altogether. The above command produced, among other findings, confirmation of the directional (one-tail) hypothesis that respondents would agree significantly more with the statement '**More legislation is needed to ensure equal rights**' (*var002*) than with the statement '**Generally speaking, men have a higher level of sexual desire than women**' (*var008*). The average values from the two seven-point scales were 2.10 for *var002* and 4.76 for *var008*. Because of the scoring method used, a low value indicates more agreement than a high value. The data from the 50 respondents yielded a *t*-value of 6.62 which corresponds to a one-tailed probability of less than 0.001. The difference is therefore in the direction predicted, and is highly significant.

The output which revealed this finding took the following form:

```
Paired samples t-test:   VAR002    New legislation needed
                         VAR008    Men more sexy

Variable      Number                  Standard    Standard
            of Cases       Mean       Deviation    Error

VAR002          50         2.1000       1.644       .233
VAR008          50         4.7600       1.890       .267


(Difference) Standard   Standard        2-Tail    t      Degrees of 2-Tail
     Mean    Deviation    Error    Corr. Prob.  Value     Freedom    Prob.

   -2.6600      2.840      .402    -.288  .043  -6.62        49       .000
```

The default output for the paired *t*-test procedure (as above) displays the mean difference of the pairs together with the standard deviation and standard error of this value. A correlation coefficient is calculated between the two variables and is printed together with its two-tail probability value. The output then includes the *t*-value, the degrees of freedom and the two-tailed probability of the difference between the means of the two variables. Cases with missing values for either variable are not included in the calculation. The output also includes the variable labels for the two variables under examination.

The **OPTIONS** available for the independent *t*-test are also available for the paired *t*-test.

OPTION 1 specifies that missing values should be *included* in the analysis.

OPTION 2 specifies listwise deletion of cases with missing data (see the **OPTIONS** section for the independent *t*-test, p. 76).

OPTION 3 specifies that variable labels should *not* be included on the output.

There is also an additional option, available only for the paired *t*-test. If **OPTION 5** is specified (there is no **OPTION 4**) then a special pairing operation is brought into action. If this is used, the command:

```
SPSS/PC: t-test pairs= var001 to var005 with var006 to var010
       : /options = 5.
```

leads not to 25 comparisons (5×5) but to *five* only. The first variable before the keyword **WITH** is compared to the first variable after **WITH**, the second before is compared with the second after, etc., until finally the last (fifth) variable before **WITH** (*var005*) is compared with the final (fifth) variable after **WITH** (*var010*).

No additional **STATISTICS** are available for the **T-TEST** procedure.

There is no SPSS/PC+ limit on the number of variables which can be compared with a paired *t*-test procedure command (there is, however a limit on the number of variables in the active file; there can be no more than 200 variables).

8 Transforming Variables; Selecting and Sorting Cases

Once a file containing the raw data has been established, the user may wish to transform the data in certain ways. Thus the coding system originally used for a variable may be inappropriate for certain types of analysis. For example, if a variable *age* has been coded in terms of years it will not be possible to directly compare a young group with a middle-aged or old group. With a **RECODE** command, however, the existing data can be re-classified so that *age* (in years) is recoded into, say, 'young' – up to 25 years; 'middle' – 26 to 50 years; and 'old' – over 50 years. Such a recoding would produce three age groups rather the original groupings (one for each separate age).

Sometimes the user may wish to create new composite variables (say, the sum of *var001*, *var002* and *var003*), and this can be achieved by using another data transformation command, **COMPUTE**.

In this chapter we will first consider the use of the data transformation commands **RECODE** and **COMPUTE**, before discussing conditional transformations (involving the command **IF**). We will then consider how particular cases can be selected for analysis, and how the order of cases within a file may be changed using the **SORT** command.

8.1 The RECODE command

A **RECODE** command specifies variable(s) to be recoded and provides an equation (=) for transforming the old values into new values (in this order).

The following examples illustrate how data may be recoded:

```
SPSS/PC: RECODE var001 (1,2,3=1) (4=2) (5,6,7=3).
```

This transforms values from the original seven-point scale into just three values: 'low' (the old codings below 4 are now coded together as 1), 'middle' (the old 4 becomes 2) and 'high' (the old codings above 4 are now coded together as 3). The system takes note of the **RECODE** instruction but the actual process of recoding is not implemented until a subsequent procedure is requested.

If you issue the above command on the data from the gender attitudes survey, follow it with a request for a **FREQUENCIES** display:

```
SPSS/PC: frequencies var001.
```

The system will recode the variables and there will be a delay before the **FREQUENCIES** display is provided. This display will then indicate that values have been recoded so that all valid data is now within the range 1 to 3. You will notice that the missing data (coded 9) is unchanged. Because there is now no valid data coded 7 the label associated with that value ('strongly disagree') is no longer included in the display. This points to the need, after a recode, to revise the existing value labels (a new **VALUE LABEL** command can be issued in response to the SPSS/PC: prompt and a new set of labels supplied that will override any existing value labels for that variable).

The keywords **THRU**, **LOWEST** (also written as **LO**), **HIGHEST** (also written as **HI**), and **ELSE** can be used. Lists of variables can be included within the **RECODE** command and the lists can include the keyword **TO**:

```
SPSS/PC: recode salary (LO thru 5000=1)
       : (5001 thru 7500=2) (7501 thru hi=3).
```

This recodes the initial *salary* values to form just three *salary* groups. Valid values of *salary* will now be within the range 1 to 3.

```
SPSS/PC: recode var002 to var005 (LO thru 4=1) (ELSE = 2).
```

This recodes the initial values for the four named variables to form two groups for each.

The keywords **LO, HI** and **ELSE** will recode a missing value if it is within the range specified. Thus in the second example above, any missing data for *salary* (originally coded 99999) will now have the value 3 (and hence be indistinguishable from high valid salary values). In the third example above, any missing data for the variables (originally coded 9) will now have the value 2 (and hence be indistinguishable from high valid values). In the following example, however:

```
SPSS/PC: recode var004 var006 to var010 (1,2,3=1)
       : (4=2) (5,6,7=3).
```

the missing value (9) would be unaffected.

Having changed coding values you will need to change the relevant value labels (and possibly the variable label).

If the original values for a variable have been collapsed (say the values for *age*, in years, have been collapsed into just three categories) then the fuller raw data will not be accessible during the remainder of this session. There is no way of recoding the three *age* categories to regain information about the respondents' actual ages (remember that you can't unscramble eggs!). The original file will have to be **INCLUDED** again for the original values to be made available.

8.2 The COMPUTE command

The **COMPUTE** command creates new variables using the data from the original dataset. Thus:

```
SPSS/PC: COMPUTE sum  = var001 + var002.
```

This would, for each case, compute a value for the new variable *sum* by adding the values for two existing variables, *var001* and *var002*. Thus *new variables can be created without adding new data*.

Arithmetic operations allowed in **COMPUTE** commands include: addition (+); subtraction (−); multiplication (*); division (/) and exponentiation (i.e. squares, cubes, etc.)(**).

Examples:

```
SPSS/PC: COMPUTE monsal = salary/12.
```

This computes the expected monthly salary (*monsal*) by dividing the (annual) *salary* by 12.

However, this might present a problem. Remember that in a previous example *salary* was recoded to a range of 1 to 3. If this has been done with the active file then the above command would now divide the **recoded** values by 12. This would clearly not produce a direct representation of the monthly salary. To retrieve the original codings in the dataset we would need to **INCLUDE** the file again (but this will restore *all* of the original values and thus cancel the effects of any **RECODE** and **COMPUTE** commands, etc. entered since the file was last made active).

```
SPSS/PC: COMPUTE wt= (wst * 14) + wlb.
```

This creates a new variable *wt* (total weight in pounds) from two original variables, the weight in stone (*wst*) and the residual weight in pounds (*wlb*).

```
SPSS/PC: COMPUTE ht= (htf * 12) + hti.
```

This creates a new variable *ht* (total height in inches) from two original variables, the height in feet (*htf*) and the residual height in inches (*hti*). This example and the previous one show how brackets can be used to specify the order of operation. Remember that A*(B+C) is **not** equivalent to (A*B)+C, e.g. if A = 2, B = 3 and C = 4 then A*(B+C) = 14 and (A*B)+C = 10.

Numeric functions (including log functions, rounding, sines, etc.) and a number of other functions can also be used within **COMPUTE** commands (see the *SPSS/PC+ Manual*, pp. B30–B31).

If a subject has a missing value for one or more of the variables used in creating a new variable, then the new variable is also treated as missing for that subject. Since it is declared to be missing by the system (rather than by the user) it is said to be **system missing**.

Using COMPUTE to score tests

One special use of **COMPUTE** is to derive a composite score from a number of related or similar variables.

Suppose, as in the gender attitude survey, we have a 12-item personality test, and wish to derive scores on two dimensions (in this case extraversion and neuroticism). The test given to respondents in this survey was devised by Hans Eysenck (1958). From the scoring key supplied with this test, we know that to derive the extraversion and neuroticism scores we need in each case to take into account six of the twelve variables *E1* to *E12*.

A 'no' response to any relevant item **lowers** the extraversion or neuroticism score by 1 point, whereas a 'yes' response **increases** the score.

Thus one way to compute extraversion and neuroticism scores would be first to recode the original data. The 'yes' codes (1) do **not** need to be changed. The 'not sure' responses (originally coded as 2) could be re-coded to 0, and the 'no' responses (originally coded as 2) could be recoded as −1. These changes could be made using the single **RECODE** command:

```
SPSS/PC: RECODE E1 to E12 (2= 0) (3 = -1).
```

A **LIST** command could be issued to examine the effect of this transformation. Following the above **RECODE** command, the **LIST** command:

```
SPSS/PC: list id e1 to e12.
```

produced the following display:

```
ID E1 E2 E3 E4 E5 E6 E7 E8 E9 E10 E11 E12

 1  1  1 -1 -1  1 -1  1  1  0   1  -1  -1
 2  1  1 -1  0 -1  1  1  1 -1   0   1   1
 3  1  1  1  1  1  1  0 -1  1   0   1  -1
 4  1 -1  1  0 -1  1 -1 -1  0   1   1   1
 5  1 -1  0 -1  1  1  1 -1  1   1   0   1
....
47  1  1  1  1 -1  1  1  1  0   1   1   1
48  1  1  1  1  1  1  1  1 -1   0   1   1
49  1  1  1  1 -1  1  1  1 -1   1   1   1
50  1 -1 -1  0 -1  1  1 -1  1   0   1  -1

Number of cases read =    50    Number of cases listed =    50
```

The extraversion score (*extra*) was then computed from six of the recoded *E* variables:

```
SPSS/PC: compute extra = (E2 + E4 + E7 + E8 + E10 + E12).
```

and a neuroticism score (*neur*) was computed from the other six *E* variables:

```
SPSS/PC: compute neur = (E1 + E3 + E5 + E6 + E9 + E11).
```

A **LIST** command will now display the values of the new variables:

```
SPSS/PC: list id extra neur.
```

ID	EXTRA	NEUR
1	2.00	-1.00
2	4.00	0.0
3	0.0	6.00
4	-1.00	3.00
5	0.0	4.00
....		
47	6.00	3.00
48	5.00	4.00
49	6.00	2.00
50	-2.00	2.00

A **FREQUENCIES** display for the new variables *extra* and *neur* shows that the mean for *extra* is 1.860 and the standard deviation is 2.799. The mean for *neur* is 2.140 and the standard deviation is 3.064.

Using the above strategy, and issuing the **RECODE** and **COMPUTE** commands interactively, the new variables *extra* and *neur* would be created for the current session only. The computed variables could be specified in any further procedure commands as long as the session lasted, but at the end of the session these variables would no longer exist.

A further example of test scoring using **COMPUTE** involves different variables. Suppose, in examining the 10 statements which relate to the gender attitude variables *var001* to *var010* (see gender attitudes questionnaire, Appendix C), we consider that agreement with some of them would reflect a non-sexist viewpoint, but that agreement with others would reflect a sexist outlook. The data reflecting an individual's responses to such statements could be used to assess that individual's degree of sexism (it is likely that individuals will differ somewhat in their assessments of which statements reflect a sexist outlook and, in practice, the construction of a test would not depend on such subjective judgements).

If we judge items 1, 2 and 9 to be non-sexist then for these items a **high** score (indicating **disagreement**) would indicate a **high** degree of *sexism*. Items 3, 4, 5 and 8, on the other hand, might be judged to be sexist.

Agreement with such statements (indicated by a **low** coded value) should therefore **increase** the *sexism* score.

A **COMPUTE** command to combine information from both the 'positive' sexism items and the 'negative' sexism items, without the need for a **RECODE**, could take the form:

```
SPSS/PC: compute sexism = (var001 + var002 + var009 +
       : (8 - var003) + (8 - var004) + (8 - var005) +
       : (8 - var008)).
```

Here a high sexism score would be obtained by having a **low** score for variables *var003*, *var004*, *var005* and *var008*, and by having a **high** score for variables *var001*, *var002* and *var009*. Agreement with any of the sexist items (orginally coded as a **low** value) has now been been made to increase the *sexism* score substantially, through the application of the 8 − var formula to the original coding.

The new *sexism* variable would have a theoretical range from 7 to 49. The actual range for the respondents in the gender attitudes survey (as shown by a **FREQUENCIES** display) is 14 to 38. The mean *sexism* score for this population is 24.229 and the standard deviation is 5.846.

Lowering the derived *sexism* scores by seven points (i.e. one point for each item) would allow those respondents who were totally non-sexist (i.e. those who strongly agreed with each of the equality items and strongly disagreed with each of the sexist items) to obtain a zero score. This lowering of the *sexism* scores by seven points across the range could be achieved by including an additional **COMPUTE** command:

```
SPSS/PC: compute sexism = sexism - 7.
```

The theoretical range of this *sexism* score would now be 0–42. The actual range for the survey respondents was 7–31. This was revealed by a new **FREQUENCIES** display but could also have been calculated from the previous range (i.e. by subtracting seven from both the previous minimum and the previous maximum). The mean is lowered by exactly 7.00 (it is now 17.229). As we would expect, the standard deviation is unaffected (it remains at 5.846).

We could also compute a problem score by combining information from several of the *prs* variables (coded 1 for 'true' and 2 for 'false'). This could be achieved by adding one point for each of a range of items which the respondent reported as a frequent occurrence ('often bored', 'often depressed', 'often anxious', etc.). Allowing for the fact that denial of such problems has been coded as 2 ('false'), and admission of such problems as 1 ('true'), the overall *problem* score could be computed in the following way:

```
SPSS/PC: compute problem = (2 - prs01) + (2 - prs03) +
       : (2 - prs04) + (2 - prs07) + (2 - prs08) +
       : (2 - prs09) + (2 - prs10).
```

This would produce a score, for each individual case, ranging from 0 ('no problems') to 7 ('many problems'). For the gender attitudes survey data the mean value for *problem* is 2.857 and the standard deviation is 1.893.

8.3 Conditional transformations: IF

The **COMPUTE** and **RECODE** transformations described so far affect all cases in the active file. Sometimes, however, it is useful to transform a variable only for those cases in which certain conditions are fulfilled. Such transformations are known as conditional transformations and they affect any particular case only **IF** a specified logical expression is true for that case.

The examples given in this section describe a hypothetical dataset. They do not refer to the gender attitudes survey data.

We might wish to assign a value 1 for a new variable *pass* only for those cases in which an examination mark (the value of a variable *test*) reaches a criterion pass mark. A suitable conditional transformation would assign a specified value for the new variable only to those cases for which the value of *test* is **greater than** a specified criterion value. Such a command could take the form:

(a) `SPSS/PC: IF (test GT 49) pass = 1.`

GT is a relational operator which stands for greater than. The above command would assign the value 1 for the new variable *pass* to any case for which the logical expression (test **GT** 49) is true, i.e. for any case for which the *test* value (e.g. an examination mark) is more than 49. For those cases in which the logical expression was evaluated as not true (or for which it could not be evaluated) the system would assign a system missing value to *pass*.

We might wish to assign another value to *pass* for those students who had actually failed the examination. For example, we could specify that all students who obtained marks less than or equal to 49 (**LE 49**) be assigned the value 0 for the variable *pass*. This could be done in either of the following ways (the relational operator **LE** stands for less than or equal to):

(b) `IF (test LE 49) pass = 0`

or

(c) `IF (NOT test GT 49) pass = 0.`

We see here that the keyword **NOT** (a logical operator) can be used to reverse the 'true' or 'false' status of an expression.

The effect of issuing the two commands ((a) and either (b) or (c)) will be to assign either the value 0 or the value 1 to all cases with a valid *test* score.

Cases with a *test* value below 50 will be assigned a 0 for *pass*. Those with 50 or more marks will be assigned the value 1 for *pass*. The logical expression will not be able to be evaluated for any case with a missing value for *test*, and any such case will be assigned a system missing value for *pass*.

Many different relational operators are permitted in logical expressions, and their various effects are outlined in Table 8.1.

Table 8.1

Relational operators

The expression (X EQ Y) is true if X is *equal to* Y

The expression (X NE Y) is true if X is *not equal to* Y

The expression (X GT Y) is true if X is *greater than* Y

The expression (X GE Y) is true if X is *greater than or equal to* Y

The expression (X LT Y) is true if X is *less than* Y

The expression (X LE Y) is true if X is *less than or equal to* Y

Relational operators can be written within commands either in character form as above (**GE, LT,** etc.) or in symbol form (thus **EQ** could be replaced by =, **GT** could be replaced by > , etc.).

The **X** and **Y** components in the table need **not** stand for simple variables or numeric values. Either or both of the components can be complex expressions themselves. Thus:

```
SPSS/PC+ IF (((v1/v2) * v3) NE ((v4 - v5) * v8)) index = 0.
```

Here the **X** of the expression is

```
((v1/v2) * v3)
```

and the **Y** of the expression is

```
((v4 - v5) * v8)
```

and if the expression X **NE** Y (X not equal to Y) is true then the variable index is set at 0.

Parentheses (round brackets) are always required around the entire logical expression to be evaluated. The position of any inner brackets within these external parentheses determines the nature of the logical expression by changing the order of evaluation. Thus in the command:

```
SPSS/PC: IF (((v4 + v5) * v6) GT 100) index = 3.
```

for each case, the value of *v4* is added to the value of *v5* and the sum then multiplied by the value of *v6*. If this product has a value greater than 100 then the expression is true for that case and the value of *index* is therefore set to 3. This sequence of operations is **not** the same as that requested by the command:

```
SPSS/PC: IF ((v4 + (v5 * v6)) GT 100) index = 3.
```

Here, for each case, the value of *v5* is multiplied by the value of *v6* and the product then added to the value of *v4*. If the resulting value is greater than 100 then the expression is true and the value 3 is assigned, for that case, to the variable *index*.

The logical expression to be evaluated as true or not true can be more complex in terms of its logic as well as its arithmetic. Several logical relations (e.g. *v1* **GT** *v2; score1* **EQ** *score2*) can be joined within a single **IF** command by the logical operators **AND** and **OR** to form a compound expression.

For example, the final evaluation of a student's performance might depend on both a continuous course-work component and the grade obtained in an end-of-course examination. An overall pass might require **both** that a score of 50 or more be obtained in the final exam **and** that the number of continuous assessment essays passed (the variable *caep*) be more than 5 (**GT 5**). The following expression would assign the value 1 for *pass* only to those candidates who were successful on **both** criteria:

```
SPSS/PC: IF (test GT 49 AND caep GT 5) pass = 1.
```

Another useful logical operator is **OR**. If, in a variation of the above example, a pass required **either** that a score of 50 or more be obtained in the final test **or** that more than 5 continuous assessment essays be passed (*caep* **GT 5**) the following expression would assign the value 1 for *pass* to those cases which reached **either** of these criteria:

```
SPSS/PC: IF (test GT 49 OR caep GT 5) pass = 1.
```

A final variation of this example involves a complex situation. Suppose that a student is awarded a pass for a course if s/he has passed more than five assessed essays and also passed the final examination (**GT 50**) **OR** if 70 marks or more have been obtained in the examination alone. The following command could be used to identify which students had passed:

```
SPSS/PC:IF ((test GT 49 AND caep GT 5) or (test GE 70)) pass = 1.
```

The terms **AND** and **OR** may be used several times in a logical expression. Thus the expression:

```
(v1 ge 100 and v2 ge 100 and v3 ge 100 and v4 ge 100)
```

is true only if **all** of the four specified variables have values of 100 or more. The expression:

```
(v1 ge 100 or v2 ge 100 or v3 ge 100 or v4 ge 100)
```

is true if **any** of the four variables has a value of 100 or more.

If two or more relational expressions are included within a single logical expression they must all be expressed in full form, as in the following examples:

```
IF (var001 LT 3 or var001 GT 5)
IF (v1 EQ 1 OR v2 EQ 1 OR v3 EQ 1)
```

The shortened forms IF (var001 LT 3 or GT 5) and IF (v1 or v2 or v3 eq 1) are not permitted.

Although each of the examples provided has involved the creation of a new variable (*pass* or *index*), existing variables can be recoded using the **IF** command. Thus, in the expression:

```
IF (v2 lt v3) v1 = 0
```

v1 could be a variable which already exists. For any case for which the logical expression was true the existing value of *v1* would be changed to 0. An extreme form of such a transformation can be seen in the command:

```
IF (v2 lt v3) v2 = 0
```

Here *v2* appears both within the logical expression and as the target transformation variable. The effect would be to recode *v2* to 0 for each case in which the initial value of *v2* was less than the value of *v3*.

Where a subject has a missing value for a variable specified in the logical expression to be evaluated, SPSS/PC+ nevertheless attempts to evaluate the overall expression. For example, in:

```
SPSS/PC: IF (v1 EQ 1 OR v2 EQ 1 OR v3 EQ 1) score = 4.
```

a particular case may have a missing value for *v1*, but may have the value 1 for *v2* or *v3*. In such a situation the truth of the overall expression can be evaluated and the subject's *score* **would** be set to 4.

Where the expression **cannot** be logically evaluated, however, the new value cannot be assigned and the new variable (*score* in the example we are now considering) is assigned a system missing value. Thus if variables were joined by **AND**, rather than by **OR**:

```
SPSS/PC: IF (v1 EQ 1 AND v2 EQ 1 AND v3 EQ 1) score = 4.
```

then a subject with a missing value for **any** of the three variables (*v1*, *v2* or *v3*) would be assigned a system missing value for the new variable *score*. If, however, *score* already exists as a variable in a dataset then, where the truth value of the logical expression cannot be evaluated, the **original** value will be retained. With 'old' and new codings likely to be intermingled, such a situation could easily lead to some confusion. The best strategy, for most purposes, is probably to use the **IF** command only to create **new** variables.

As with the other data transformations, the use of a conditional transformation (**IF**) command may require (**after** variables have been introduced or changed by the data transformation command) additions or revisions to variable labels and value labels.

When several data transformation commands are to be issued in a single interactive session, the order in which the commands will be entered should be carefully considered.

Thus if you intend to **RECODE** *v3* at some point, and you are also going to use *v3* in a **COMPUTE** command, you must decide whether you wish to **COMPUTE** the new variable using the original coded values for *v3* or using the re-coded values. Similarly, if you are going to evaluate a logical expression involving a **computed** variable you will obviously need to create the new variable **before** you refer to it in a conditional (**IF**) command.

8.4 Selecting cases

Sometimes an analysis needs to be performed on only **some** of the cases in the current active file. The cases to be selected for inclusion in an analysis may be:

- cases which have a particular characteristic (e.g. *sex* = male)
- the first *N* cases in the file (e.g. *N* = 100, or *N* = 300)
- a random sample (e.g. a random 1 in 10 cases)

Such a selection may be required to remain in effect for the rest of the current session, or for the execution of just one procedure.

It must be emphasized that some of the selection commands discussed below will cause cases which are not selected to disappear from the active file. The only way of retrieving these lost cases will be to **INCLUDE** the original file (e.g. "gender.def") once again. But in doing this, the effects of any transformations which have previously been made (e.g. with **COM-PUTE** or **RECODE** commands) will be cancelled. In Chapter 9 we will consider how transformation commands can be inserted within the data definition file, to be executed whenever that file is **INCLUDED**. At this stage, if you have been following the text by issuing commands for operation on a real file, you must decide whether you really wish to remove cases permanently (i.e. until the end of the session) from the current active file. You may choose, instead, simply to read the instructions below, and to keep your full current file, including any transformations you have made. Some of the selection commands which follow, however, have only a temporary effect and will not lead to cases being removed from the active file. You can explore the use of these temporary commands without affecting the active file contents.

To select cases with a particular characteristic

The **SELECT IF** command can be used to select cases permanently. Thus:

```
SPSS/PC: SELECT IF (SEX EQ 1).
```

would be used to select only those cases in which the variable *sex* had the value 1. The truth-value of the logical expression would be evaluated for each case separately. Where the expression was true the case would be selected 'in'. Where the expression was not true (or could not be evaluated) the case would not be selected and would be removed permanently from the active file. Thus the above command would be used to select for subsequent analyses all the male cases, only, from a population in which males were coded 1 and females were coded 2.

As well as **EQ**, the other relational operators **NE**, **GT**, **GE**, **LT**, and **LE** can be used, as in the conditional transformation (**IF**) commands discussed above.

The **SELECT IF** command is also similar to the **IF** command in other ways. Complex conditional statements involving the logical operators **AND, OR**, and **NOT** can be used, as can numerical expressions, for example:

```
SPSS/PC: SELECT IF (WST GT 10 or HTF GT 5).
SPSS/PC: SELECT IF (var001 + var002 + var009 GT 16).
```

Brackets are used to specify the order of evaluation, e.g.

```
SPSS/PC: SELECT IF ((v1 ** 2 - 100) * (v2 - v3)  GT 100).
```

In the command:

```
SPSS/PC: SELECT IF (var001 le var002 or var003 gt var001).
```

a case would be selected for analysis if, and only if, the value of *var001* was less than, or equal to, the value of *var002* **OR** if the value of *var003* was greater than the value of *var001* (or if **both** of these conditions were fulfilled).

A **SELECT IF** command affects all subsequent procedures (and thus has a permanent effect). Examine the following example from the gender attitudes survey:

```
(a) SPSS/PC: frequencies var001.
    SPSS/PC: SELECT IF (sex eq 1).
    SPSS/PC: frequencies var002.
    SPSS/PC: crosstabs var004 by var005.
```

Assuming that the active file initially includes all of the cases created by the **INCLUDE** "gender.def" command, the first procedure (frequencies *var001*) would be performed on **all** 50 cases.

After the **SELECT IF** command, however, the subsequent procedures (both **FREQUENCIES**, using *var002*, and **CROSSTABS**) would be performed only with those cases selected to remain in the active file (in this example, only the 22 male cases).

Because only the male cases now remain in the active file, any subsequent procedure will also operate only on the 22 male cases. To regain the full complement of 50 cases (to bring back the females) we would need to **INCLUDE** the original file once more.

Because a **SELECT IF** command affects all subsequent procedures, multiple uses of the command will lead to a **progressive** selection. Thus, in the example:

(b) SPSS/PC: SELECT IF (sex eq 1).
    SPSS/PC: frequencies var002.
    SPSS/PC: SELECT IF (var001 gt 3).
    SPSS/PC: crosstabs var004 by var005.

the **CROSSTABS** procedure will be performed only on those cases which have the value for *sex* which is specified in the first **SELECT IF** command (22 cases) **and** which have a value greater than 3 for *var001* (this reduces the number of cases in the active file to just 11).

It is possible to specify that the selection of cases should apply only to the procedure which immediately follows, i.e. that a temporary selection should be made. In such circumstances the **SELECT IF** command is replaced by **PROCESS IF**. Thus if we substitute **PROCESS IF** in the above example (a), giving:

(c) SPSS/PC: frequencies var001.
    SPSS/PC: PROCESS IF (sex eq 1).
    SPSS/PC: frequencies var002.
    SPSS/PC: crosstabs var004 by var005.

then (assuming that we started with the full complement of 50 cases) the first command (frequencies var001) would operate on all cases. The procedure immediately following the **PROCESS IF** command (the **FREQUENCIES** command specifying the variable *var002*) would be performed using only the selected (22) cases. For any further procedures specified, however, **all** of the 50 cases in the original file would be included in the analysis. Thus the **CROSSTABS** procedure would be performed using **all** valid cases in the original population.

Thus with a series of suitable **PROCESS IF** commands, each followed by a particular procedure, different samples of the population could be analyzed successively. After each analysis the complete original set of cases would again be available.

Although the use of any of the relational operators (**EQ, GT**, etc.) is permitted within **PROCESS IF** commands, the logical operators (**AND, NOT** and **OR**) are **not** permitted.

To select the first N cases in a sample

The **SELECT IF** and **PROCESS IF** commands are used to select cases which conform to specified criteria. Sometimes, however, the researcher may wish merely to reduce the number of cases rather than to select cases which are of a particular type. One way of achieving such a size reduction is to specify the **number (N)** of cases to be selected. For example:

```
SPSS/PC: N 40.
```

This specifies that future procedures should operate only on the first 40 cases in the file. As with the other transformation commands, however, the N command will not in itself produce an immediate effect. The system will note the request but not execute the selection until the next procedure command is issued to the system.

The question of permanence is rather complex in the case of **N** commands. The first **N** command creates a working dataset and, once a procedure or transformation has been used with this, the number of cases cannot be increased during the rest of the session. The **first N** command, therefore, can be regarded as permanent, and any further **N** commands issued during the session must specify a lower number. Subsequent **N** commands, however, do not affect the working dataset, so that with, say, a working dataset of 40, a second command **N 20** can be executed and then followed by a third command **N 35** (but **not N 45** because that would exceed the size of the working dataset). Thus any **N** commands issued subsequent to the first one may be regarded as temporary.

The following sequence of commands may make this clearer:

Command	Cases available for procedures
include "gender.def".	(50)
N 40.	(50 – procedure not yet executed)
frequencies sex.	(40 – now the working dataset)
N 20.	(40)
frequencies fac.	(40)
N 25.	(40)
frequencies yr.	(40)

There is, however, a further complication. If a transformation is requested (by, say, a **RECODE** command) and activated (when the data are read for the execution of a procedure command) after **any N** command then the effect of that **N** command will be permanent. There is a good reason for this. If a second **N** command **N 30** was followed by a **RECODE**, and **N 35** was then allowed to increase the number of cases, the first 30 cases processed subsequently would have the recoded values and the last five cases would have the original values, producing a confused tangle of results.

One last point. If an **N** command immediately follows, **or is immediately followed by** a **SELECT IF**, **PROCESS IF** or **SAMPLE** command (see below), then the **N** command is executed **after** that other command. Thus even if the command **SELECT IF** occurs after **N 10**, the conditional selection is made first and **then** the first 10 **SELECT**ed cases are used for subsequent analyses.

To select a random sample

An alternative strategy for selection (which is always temporary and therefore always affects only the next occurring procedure) involves specifying a **proportion** of cases to be **randomly** selected from the total file.

This strategy involves using the **SAMPLE** command to select a random sample from the total population of cases in the active file. The command can be used in either of the following ways:

```
(a) SPSS/PC:    SAMPLE  .10.
(b) SPSS/PC:    SAMPLE  100 from 1010.
```

In example (a) the user has specified that a 1 in 10 random sample is to be taken. The system will select a random 10 per cent of cases (approximately).

In example (b) the user has specified how many cases, s, (in this instance 100) are to be randomly sampled from the total file population (which, in this instance, the user is claiming to be , S, 1010). If a mistake is made in estimating the size of the total file population, adjustments are automatically made to the sampling. If the true population size is **less** than **S**, then the number sampled is proportionately less (thus if the true size were 909, the number sampled would be 90). If the true population size is **more** than that specified, then a full-sized sample (containing s cases) is taken, but that sample is taken randomly from only the first S cases.

It must be emphasized that the process of sampling, following a **SAMPLE** command, is a random one and that even if the same number (or proportion) of cases is sampled on a number of occasions the particular **cases** sampled will be different. In an exercise to explore the effect of this, file "gender.def" was made active and exactly the same pair of commands:

```
SPSS/PC:   sample 25 from 50.
SPSS/PC:   frequencies var001/ statistics = mean.
```

was issued 10 times. The mean values of *var001* for the 25 cases selected on each occasion were noted, and these ranged from 2.480 to 3.560 (the mean for all 49 valid cases in the file is 3.102).

8.5 Sorting cases

The order in which cases occur within the file can be changed by a **SORT** command. This arranges cases in ascending or descending order of their value on any variable. **SORT** is a procedure command and therefore, unlike transformations commands, it produces an immediate response.

The default order for a **SORT** command is **ascending**. To reverse this we must specify a descending sort (**D**). Thus, with the variable *age* coded in terms of years:

```
SPSS/PC: SORT by age.
```

will sort the cases in ascending age order (i.e. from the youngest to the oldest).

A descending sort would place the oldest above the youngest:

```
SPSS/PC: SORT by age (D).
```

To observe the effect of a sort, the **LIST** command can be used. Thus if, after issuing the above command, **LIST** is requested, the values for all of the variables will be listed case by case. Using this example, it can be observed from the **LIST** display that the *id* variable is out of order as the cases are displayed, but that cases are now presented in descending order of *age*.

If a **SORT** command has been issued **interactively** then, as long as the file remains active, the original order can be restored. If the file originally contained cases in the order of an *id* variable, of course, the initial order can be restored (for any cases which remain in the active file) by issuing the command:

```
SPSS/PC: sort by id.
```

In other situations, advantage can be taken of the fact that SPSS/PC+ assigns a system variable *$casenem* to each case (in the order it occurs within the file) as a file is made active. The original order can be restored at any time in the session by using *$casenum* as the sorting key:

```
SPSS/PC: sort by $casenum.
```

Sorts involving multiple criteria (or sort keys) can be specified. Thus:

```
SPSS/PC: SORT by age fac (D).
```

would produce a sort in which cases were arranged first in descending order of age. If two or more cases shared the same age, then the next sorting criterion *fac*, the respondent's faculty, would be used to order cases **within the same age group**. Again the order would be in terms of descending value (in this case social studies, coded 3, then science, coded 2, then arts, coded 1). Notice from this example that the single **D** at the end of the command suffices to specify a descending sort for **both** of the variables *age* and *fac*.

An ascending sort can be specified by using **A**, so that the sorting command can specify any mixture of ascending and descending sorts. Thus, resulting from the following command:

```
SPSS/PC: SORT by fac (A) age (D) salary (A).
```

the initial, or major, sort would be an **ascending** sort in terms of *fac* (arts, then science, then social studies). Cases within the same faculty would then be ordered by descending *age*. Cases with both the same *fac* coding and the same *age* would be further sorted by ascending *salary*.

Precisely the same order of cases could be achieved by issuing three separate **SORT** statements but special care must be taken with the order in which they are issued. The major sort must come **last**, so the three **SORT** commands would to achieve the same result would be SORT by *salary* (a), **then** SORT by *age* (d), and **finally** SORT by *fac* (a). This illustrates the fact that any **SORT** starts with the cases in their current order (not in the original order) and does the minimal amount of re-ordering required to satisfy the **SORT** request.

Combining selections and sorts

Sorting and selection commands can be combined to produce many different kinds of samples for analysis. The following examples illustrate some effects of combining such commands. In each case the commands specified were issued immediately following the inclusion of the **full** "gender.def" file (i.e. **50** cases).

(a)   SPSS/PC: sort by sex (D).
        SPSS/PC: sample 14 from 28.
  (i)   SPSS/PC: frequencies var001.
  (ii) SPSS/PC: frequencies var002.

Since the variable *sex* is coded 1 for males and 2 for females, the **descending** sort of cases by *sex* places the 28 females in the total sample at the beginning of the active file. The temporary **SAMPLE** command samples 14 from the first 28 cases (here, all of the female cases) for analysis (i). All 50 cases are analyzed by command (ii).

(b)   SPSS/PC: select if (sex EQ 2).
        SPSS/PC: sample 14 from 28.
  (i)   SPSS/PC: frequencies var001.
  (ii) SPSS/PC: frequencies var002.

Here the **SELECT** command remains in effect for all subsequent procedures whereas the **SAMPLE** command is temporary. Analysis (i) therefore

involves 14 (female) cases, and analysis (ii) involves all 28 females (only –
the males have been permanently excluded by the **SELECT IF** command).

```
(c)  SPSS/PC: process if (sex EQ 2).
     SPSS/PC: sample .5.
(i)  SPSS/PC: frequencies var001.
(ii) SPSS/PC: frequencies var002.
```

Here the **PROCESS IF** command leads to the temporary selection of the 28
female cases. **SAMPLE** leads to the inclusion of approximately 50 per cent
of these. Thus about 14 cases will be analyzed as a result of command (i).
At stage (ii), however, neither of the temporary selections is operative,
and data from all 50 cases are included in the analysis.

In this chapter we have considered how values may be recoded and how
new variables may be created, and we have examined the ways in which
cases may be selected, sampled and sorted.

9 Command Files and System Files

By now you should have explored the use of some or all of the procedures **FREQUENCY, CROSSTABS, MEANS** and **T-TEST**. You should also have explored the ways in which variables can be recoded, and new variables computed, and the ways in which cases can be selected or re-ordered. You may have been using either a data definition file containing 'in-line' data or a data definition file which includes a reference to a separate data file.

In this chapter we consider **command files** and **system files**. We already have some experience of one kind of command file – the data definition file. System files, however, have not yet been considered.

Let us first recall the story so far. The editor program **REVIEW** has been used to create data files and data definition files for use by SPSS/PC+. For example, where the data are contained in a separate data file:

- Data have been entered and the dataset then written to disk as a **data file**.
- A **data definition file** has then been written. This lists the variables represented within a specified data file, with formatting information, variable labels and value labels. Missing values are assigned to variables and the complete data definition file then written to disk.

After a data definition file has been written to the disk it has been named in an **INCLUDE** command to become the active file, e.g.

```
SPSS/PC: include "gender.def".
```

Once the active file has been created, the data within it can be processed. Various types of command can be issued interactively (i.e. in response to the SPSS/PC: prompt or the continuation prompt :).

This is a good time to recapitulate on the various types of SPSS/PC+ command.

9.1 Types of command

Commands can be divided into three different groups:

(a) **Operation commands** provide information or change parameters of the SPSS/PC+ operation – they do not transform or analyze data. This group of commands includes **HELP, DISPLAY, SHOW, SET, REVIEW, INCLUDE,** and **FINISH**. They are executed immediately after the command is issued.

(b) **Data definition and manipulation commands** provide information to the system about the dataset or about how it should later be used. Such commands may specify the file in which data is to be found, and describe how the dataset is arranged; they may provide information about variable and value labels, transformations, and the selection of cases. So far we have met **DATA LIST, VARIABLE LABELS, VALUE LABELS, MISSING VALUES, RECODE, COMPUTE, SELECT IF, PROCESS IF, N,** and **SAMPLE**. These commands are not executed immediately but are noted by the system and executed when a subsequent procedure command is entered.

(c) **Procedure commands** read the data and, usually, perform analyses. The initial procedure command (for example, **FREQUENCIES**) following a 'read data' command (for example, **DATA LIST**) creates an active file as defined by the 'read data' command. Subsequent procedure commands read the data from the current active file. When such a procedure command is entered, any data definition or manipulation commands which have been noted by the system but not yet put into effect are first executed, and the procedure itself is then executed. We have so far encountered the procedure commands **LIST** (which displays data values) and **SORT** (which re-orders cases in the file), neither of which analyzes data, and the statistical procedure commands **FREQUENCIES, CROSSTABS, MEANS, T-TEST**.

So far most of these commands have been entered interactively, although some have been written into the data definition file (**DATA LIST, VARIABLE LABELS, VALUE LABELS, MISSING VALUES**). The commands which have been entered interactively, however, can also be written into files. These are known as **command files** (and thus a data definition file is a special type of command file).

9.2 Command files

When an interactive session is brought to an end (with the command **FINISH**), the original data definition file remains unaltered on the disk, ready to be retrieved at a later date for further analysis. However, any transformations (including recodes and new computed variables) are lost and the relevant commands will need to be entered again if the transformations are to be re-introduced.

An active file may continue to be worked on interactively, no matter how complex the transformations and analyses might become, but such a strategy is by no means efficient. The solution to this problem is to create a command file which includes such transformation (and perhaps other) commands. These commands are then executed by the system, one by one, as if they had been entered interactively from the keyboard. Thus a whole series of operation, transformation and procedure commands can be written as a file (or into an existing data definition file) and the sequence of commands then executed automatically by the system. Processing which is controlled by a command file, rather than by interactive input, is referred to as batch processing.

The user may either add further commands to the data definition file or create a new command file.

Command files, like the data files and data definition files we have already encountered, can be created or edited with an editor program. In the following discussion it will be assumed that the SPSS/PC+ editor **REVIEW** is used. Those who prefer to use an alternative editor should refer to Appendix B.

In theory a command file can contain any command which can be issued interactively (including operation commands, data transformation and manipulation commands, and procedure commands). In practice, however, some types of command are unlikely to be inserted within a command file. **HELP**, for example, is generally issued only interactively, when a problem arises or when 'on-line' information is needed. A **HELP** command may be useful when the user is writing a command file, but it is unlikely to be useful if inserted within the file itself.

Similarly, **REVIEW** is unlikely to be inserted into a command file.

To illustrate how a complex command file might be created from an existing data definition file, we can add some **RECODE** commands and some **COMPUTE** commands to the data definition file "gender.def" (from the gender attitudes survey). In Chapter 8 some suggestions were provided for data transformations to be performed on this file. In particular, suggestions were given for:

- recoding *salary* into just three values (p. 81)
- computing the variables *wt* and *ht* (p. 82)
- scoring a test for *extra* and *neur* (pp. 83–4)
- computing a *sexism* score (pp. 84–5)
- computing a *problem* score (p. 85–6)

Because the relevant commands were issued interactively, at the end of the session, the recoded variables and the new computed variables were lost. Such transformations could be made a permanent feature of the file by adding the relevant transformation commands directly into the data definition file.

To do this, we would first use **REVIEW** to bring the original data definition file forward for editing:

```
SPSS/PC: review "gender.def".
```

A copy of the contents of this file would now be displayed on the screen, with the cursor at the beginning of the last line. Moving the cursor to the end of this line and pressing <RETURN> would provide a new blank line. To request the recodes and the new computed variables in the list above we could then add the following lines:

```
recode salary (0 thru 5000=1) (5001 thru 7500=2) (7501 thru 99990=3).
compute wt= (wst * 14) + wlb.
compute ht= (htf * 12) + hti.
recode E1 to E12 (3= -1) (2= 0).
compute extra = (E2 + E4 + E7 + E8 + E10 + E12).
compute neur = (E1 + E3 + E5 + E6 + E9 + E11).
compute sexism = (var001 + var002 + var009 + (8 - var003)
(8 - var004) + (8 - var005) + (8 - var008)).
compute sexism = sexism - 7.
compute problem = (2 - prs01) + (2 - prs03) + (2 - prs04) +
(2 - prs07) + (2 - prs08) + (2 - prs09) + (2 - prs10).
```

To bring the labelling up to date and in line with these transformations, we should also add labelling revisions. Thus:

```
Variable labels salary "Recoded salary"
/ wt "Weight in pounds" / ht "Height in inches"
/ extra "Extraversion score" / neur "Neuroticism score"
/ sexism "Sexism score 0 - 42 from attitude variables"
/ problem "Problem score 0 - 7 from prs variables".
Value labels salary 1 "0 thru 5000" 2 "5001 thru 7500"
3 "7501 thru 99990 (missing NOT included)"
/ E1 to E12 1 "yes" 0 "not sure" -1 "no".
```

Procedure commands could also be added into the file. For example, we could request **FREQUENCIES** for some of the newly created variables by adding:

```
frequencies extra neur problem sexism
/barchart
/statistics = mean stddev mode range.
```

The edited file could then be written to disk. There are two ways of doing this. We could:

(a) (i) make a block of the whole file, from the initial **DATA LIST** command to the final line, using the **REVIEW** <F7> function.

(ii) then write the block to disk (<F9> in **REVIEW**), supplying a suitable filename (e.g. "gender.sps" – the extension '.SPS' is often used for SPSS/PC+ command files, to distinguish them from **DOS** '.COM' files).

or

(b) write the file to disk (< ↑ F9> in **REVIEW**), supplying a suitable filename (e.g. "gender.sps").

The command file "gender.sps" will now be available on the disk. Because a new name has been specified for this file, it will not have over-written any existing file. In particular, the original data definition file ("gender.def") will still be available.

To execute the series of commands contained within the command file we would first exit from **REVIEW** (using <aF10>) and then respond to the SPSS/PC+ prompt with an **INCLUDE** command specifying the new command file. For example,

```
SPSS/PC: include "gender.sps".
```

As the command file is being **INCLUDED**, the transformation commands will be noted by the system. These will be executed immediately before a subsequent procedure, and any procedures specified within the command file will then be performed.

If any problem emerges during the inclusion process, an error message will be displayed. The command file can then be corrected by issuing a command such as

```
SPSS/PC: review "gender.sps".
```

and editing out any errors in the normal way.

When a command file has been successfully included, and after any transformations and specified procedures have been performed, the file will remain as the active file and the SPSS/PC: prompt will be presented.

If further analyses are required on the same dataset there are several ways of proceeding:

- the current active file (which contains transformations, etc.) can be processed interactively
- the existing command file (based on the original data definition file) can be re-edited with **REVIEW**
- a new command file can be created with **REVIEW**

If the existing command file is to be revised (the second possibility above), then some of the existing procedure commands may need to be deleted (using the line delete function <ᴧ F4>), to prevent the system from repeating analyses which have already been executed. New procedure commands could then be added. When the revised command file has been written to disk, and an exit made from **REVIEW**, the revised command file would need to be **INCLUDED** in response to the SPSS/PC: prompt.

Alternatively, a new command file can be written. Such a file may be prepared in one session for use in a later SPSS/PC+ session, or may be written while the current active file is still available (since using **REVIEW** does not affect the active file). If this is done, then when the new command file is **INCLUDED** the commands it contains will operate immediately on the active file.

If a new command file is to operate on a dataset which is not currently the basis of the active file, an appropriate **INCLUDE** command can be placed at the beginning. When the command file is then executed the relevant data definition file will be included before subsequent commands are issued from within the command file.

Thus, the following command file merely refers to the relevant data definition file with an initial **INCLUDE** command:

```
include "gender.def".
recode E1 to E12 (2= 0) (3 = -1).
compute extra = (E2 + E4 + E7 + E8 + E10 + E12).
compute neur = (E1 + E3 + E5 + E6 + E9 + E11).
frequencies extra neur
/statistics = mean stddev mode range.
```

This is an example of a complete command file. It could be written to disk as, say, "extra.sps". At any later stage, the interactive command:

```
SPSS/PC: include "extra.sps".
```

would be sufficient to produce a display of the frequencies requested by the command file.

The first line of this command file will lead to the inclusion of "gender-.def", which in turn contains a reference to the relevant **data file** ("gender-.dat"). The active file will therefore include information which has been obtained from three relevant files although only one of these ("extra.sps") has been directly specified in the interactive **INCLUDE** command.

Command files are especially useful when a similar sequence of procedures is to be performed a number of times (e.g. with different variables, different **OPTIONS**, or different **STATISTICS**). Rather than issuing a series of similar commands repeatedly in the interactive mode, a command file could be written once and then edited a number of times to produce the desired variations on the initial analysis.

Operation commands (**SET**, etc.) can also be added into command files. Thus, executing the command file:

```
include "gender.def".
set printer on.
frequencies var001.
set printer off.
frequencies var002.
```

will produce a screen display of both **FREQUENCIES** commands. Only for the first of these (frequencies *var001*), however, will there be a printed record.

A separate command file can also be set up to issue a batch of operation commands. Suppose that we generally work interactively within a session, exploring aspects of the dataset by referring to screen displays, but that at the end of a session we generally write a command file containing any analyses for which we require a printed record. We would probably wish to obtain this record without the pause between pages (accompanied by the **MORE** message). Suppose also that we generally print in wide format and that we always request that each separate page of output be printed on a new physical page (by using the **EJECT** facility). Rather than issuing the necessary series of instructions interactively each time we switch into a printing phase of a session, we could create a command file containing all of the necessary operation commands:

```
set more off.
set printer on.
set eject on.
set width wide.
```

This could be written to disk as a file named "printon.sps" to be **INCLUDED** whenever we wish to switch into a printing phase.

It might also be useful to create another such file ("printof.sps") to reverse these operations. When this is **INCLUDED** the operation parameters will be reset so that they are more suitable for the display phase of the session.

Several such command files, each containing a different batch of useful operation commands, might be created. Until deleted, they would remain on the disk. Since they make no reference to any specific file, they would be available as general 'tools' when the user is working with any active file.

If such a file of operation commands is created and written to the disk with the name "spssprof.ini" it will be automatically executed each time that SPSS/PC+ is run. Thus if you were working continually on the "gender.sps" file, and preferred the beep off and the printer on, the following "spssprof.ini" file would be useful:

```
set printer on.
set beep off.
include "gender.sps".
```

It is assumed here that you are working from within your own **DOS** directory. In this case the "spssprof.ini" file would be automatically stored in this directory. This means that the particular "spssprof.ini" commands will be issued when *you* use SPSS/PC+ but not when other people use the system from their own directories.

As your project reaches a new phase, or as you change from one project to another, you will want to change the commands within this file. You can erase the file at any time (from **DOS**, using the erase command), you can edit the existing "spssprof.ini" file with **REVIEW**, or you can use **REVIEW** to create another file of the same name (and this new file, when saved, will overwrite the old version).

9.3 System files

Another type of file which may increase the efficiency of operating with SPSS/PC+ is the **system file**. This is a complete record of the active file in its current state (i.e. with any selections, transformations, re-sorts, labels, etc. remaining in effect).

Within a system file, the information is stored in a special SPSS/PC+ format and written in binary form (i.e. as 'bits' of information) rather than in ASCII form (i.e. as alphanumeric characters). The advantage of working with a system file is that processing time is significantly reduced. A disadvantage, however, is that *system files cannot be edited*.

The commands for writing a system file to disk, and for later making such a file active, are not the same as those used for command files. System files are written to disk with a **SAVE** command (save outfile = **filename**) and are subsequently read with a **GET** command (not with an **INCLUDE** command).

Thus, to save the current active file, we would issue the command:

```
SPSS/PC: save outfile= "gender.sys".
```

If the **SAVE** command is used without specifying a name, the default name "SPSS.SYS" is used. Thus:

```
SPSS/PC: save.
```

will produce a system file named "SPSS.SYS". If a file with that name already exists it will be overwritten.

A system file will contain the current contents of the active file, including data and data dictionary information. Thus variables which have been recoded, for example, will be represented in terms of their new values (only). If a permanent selection command has been issued, only the remaining cases will be represented in the system file. Thus, for example, if the complete "gender.def" file had been included, and the female cases then selected for analysis (with **SELECT IF**), only the data from the female respondents would remain in the active file. A **SAVE** command issued at this stage would therefore lead to the creation of a system file containing only the information relating to the female respondents. Similarly, if a

SORT command has been issued, then when the file is **SAVED** the cases will be stored in their re-ordered sequence.

Thus special care is needed when working with an active file which is subsequently to be **SAVED**. The commands **RECODE**, **COMPUTE**, **IF**, **SELECT IF**, **N**, and **SORT** have a permanent effect on the contents of the active file (and will thus affect the contents of any system file subsequently created from it).

Selection commands which have only a temporary effect, however (**PROCESS IF** and **SAMPLE**), do not change the active file and therefore have no effect on any subsequently created system file.

It is possible to issue a **DROP** subcommand within the **SAVE** command to request that not all of the variables in the current active file be included in the system file which is to be written to disk. The **DROP** subcommand should list those variables which are **not** to be included in the system file. The **TO** convention can be used, for example

```
SPSS/PC: save outfile= "gender.sys" / drop fac yr prs01 to prs10.
```

The system file will now contain all of the variables in the current active file except those named on the **DROP** subcommand. But the dropped variables will remain in the active file, for the use of **SAVE** does not affect the active file.

In addition to the user-supplied information (data, labels, missing values, etc.) the system file will also contain extra information supplied by the system – system variables. One of these, *$casenum* was introduced when we considered the use of the **SORT** command in Chapter 8. *$casenum* is an identifier number given to cases (in the order in which they occur within the dataset) as they are read into the active file. Even after **SORT** commands have been issued, the *$casenum* for each case remains the same. The value for *$casenum* is stored with the other data for a case when it is **SAVED** as part of a system file. Thus it will be possible (by using a 'sort by *$casenum*' command) to re-order cases which have been read from a system file so that they are arranged in the same order as when the source file was first made active.

Once a system file has been written it can be read. Because of the special form in which such files are written, however, the **INCLUDE** command **cannot** be used to read system files into SPSS/PC+. Having been created using a **SAVE** command, system files must be read into SPSS/PC+ with a **GET** command.

GET, without a specified filename, will read the default system file ("SPSS.SYS"). With a file name specified it will read that system file, e.g.

```
SPSS/PC: get file= "gender.sys".
```

This would lead to the dictionary of the system file "gender.sys" being read

(**GET** is not a procedure command – the data would be read when a procedure was subsequently issued).

A **GET** command can also be issued from within a command file. Thus if the system file "gender.sys" has previously been **SAVED**, we can **GET** it with the command file:

```
get file= "gender.sys".
frequencies var001 to var003
/statistics= mean stddev mode range.
```

As with **SAVE**, a **DROP** subcommand can be used with **GET**. Thus:

```
SPSS/PC: get file= "gender.sys" / drop var001 to var010.
```

The addition of a **DROP** subcommand has the added effect of creating an active file. If no **DROP** subcommand is used, then no active file is created until after a transformation or selection command has been issued. If only procedure commands are issued (and if no **DROP** subcommand has been included in the **GET** command) then the system reads the data from the system file each time it executes a procedure.

When a system file has been read it can be worked with interactively, so that by recoding and computing variables, changing labels, and sorting and selecting cases, a limited revision of the contents of the file can be achieved.

Provided that an active file has been created, the **SAVE** command can then be used to write a new system file. This might retain the original name or, alternatively, after the active file has been tailored interactively, it could be **SAVED** using a **new** name, for example

```
SPSS/PC: SAVE file= "gentrans.sys".
```

In this case (if "gentrans.sys" were not the name of the original system file) there would now be **two** relevant system files on the disk: the original (**without** the recent changes), and a new one, "gentrans.sys" (**with** the recent changes). Either of these could be called up in future with the appropriate **GET** command:

```
SPSS/PC: GET file= "gentrans.sys".
```

or

```
SPSS/PC: GET file= "gender.sys".
```

Sometimes the user may wish to keep several different versions of the same file on disk, each with its different selection of cases, transformations, labels, etc. One or another of these can then be made the active file (with **GET**) as it is required in various stages of the project analysis.

It is clearly important to give meaningful names to files and to make a note of the state of the data in any particular file. A reminder of the names and labels of the variables in the file can be obtained by issuing the **DISPLAY** command. Fuller information can be obtained by naming particular variables on the **DISPLAY** command or by using **DISPLAY ALL**.

Whatever system files may be created, it is essential to retain on the disk the full details of the data in the form of the data file and the data definition file. As long as these files are not deleted, a fully editable version of the data and data dictionary will be available.

Indeed, although procedures will be more quickly and efficiently executed using system files, the user may prefer to work with the types of file which may be slower in operation but are editable.

Thus, instead of saving several different system files for the same study, a number of command files can be written, each specifying a different set of selections, recodes, transformations and procedures for the same basic dataset.

Other types of file will be introduced in later chapters. We will now turn our attention to further statistical analyses, however, and make use of the command file which has been developed to transform variables from the gender attitudes survey.

10   Statistical Analysis III: CORRELATION and PLOT

In this chapter we consider the use of two further procedure commands, **CORRELATION** and **PLOT**.

10.1 The CORRELATION procedure

The **CORRELATION** procedure calculates the correlations between variables using the **Pearson product-moment** correlation formula. The procedure will tell us the degree to which pairs of variables are correlated.

A **positive** correlation means that high values of one variable (**A**) tend to be associated with high values of the other variable (**B**) and that low values of **A** tend to be associated with low values of **B**. We would expect, for example, that for the subjects in the gender attitude survey, the *problem* variable would be positively correlated with the *neur* (neuroticism) variable, and that *ht* (height) would be positively correlated with *wt* (weight). Incidentally, it would make no difference to the value of a correlation between measures of height and weight whether height was recorded in inches or centimetres, or whether the weight was entered as grams, pounds or kilograms.

A **negative** correlation means that high values of **A** tend to be associated with low values of **B** (and that low values of **A** tend to be associated with high values of **B**). We might expect, for example, that *var001* (the degree of agreement with the statement: **'Women are equal to men in every way'**) would be negatively correlated with *var004* (the degree of agreement with the statement: **'There are many activities in which men are clearly superior to women'**).

The value of a correlation is represented by the **correlation coefficient**. Negative correlations range from -1 to 0 (e.g. -0.86 is a strong negative association, -0.16 is a weak negative association). Positive correlations range from 0 to $+1$ (e.g. 0.78 represents a strong positive association, and 0.04 represents a very weak positive association). The SPSS/PC+ **CORRELATION** procedure displays correlation coefficients to four decimal places (e.g. -0.6354, 0.4653).

CORRELATION also displays the **significance** of each correlation coefficient calculated. It assumes that the user has hypothesized either a positive or a negative association and therefore normally displays the **one-tailed** probability of the correlation. If a directional hypothesis has not been made, the significance of the **two-tailed** probability can be requested by **OPTION 3**.

A single asterisk ($\star$) indicates a one-tailed probability of less than 0.01. This means that a correlation of this level (or greater) would occur by chance (e.g. with repeated analyses of different subsets of random data) less than once in every 100 analyses.

A double asterisk ($\star\star$) indicates a one-tailed probability of less than 0.001. This means that a correlation of this level (or greater) would occur by chance less than once in every 1000 analyses.

The Pearson product-moment correlation makes certain assumptions about the kind of data used. A statistics textbook will provide details of these assumptions. The product-moment correlation is used where the data (for **both** variables) are true numbers (to correlate codes or ranks, other forms of analysis are used).

Using the CORRELATION procedure

The command for a **CORRELATION** procedure takes the form:

```
SPSS/PC: correlation var001 with var002.
```

The normal listing rules and conventions hold, and multiple lists can be included in a single **CORRELATION** command. Thus:

```
SPSS/PC: correlation var003 to var006 with var007 var008
       : /var001 to var005 with var009 var010.
```

This requests 18 correlations altogether (8 are requested in the first subcommand and 10 in the second). The correlation values are displayed in the form of two rectangular matrices (one for each subcommand), with variables **before** the keyword **WITH** forming the **rows** (and those **after** forming the **columns**).

If the keyword **WITH** is omitted then correlations are performed between all pairs of the variables specified. Thus:

```
SPSS/PC: CORRELATION var003 to var006.
```

would produce a square matrix of (4×4) 16 correlation coefficients. However, five of these would be replications (e.g. *var003* would be correlated with *var004* and *var004* would be correlated with *var003*, producing the same result on both occasions). Also, each of the four variables would be correlated with itself, producing a correlation coefficient of exactly 1 (displayed by **CORRELATION** as 1.0000). Of the 16 coefficients in the matrix, therefore, only five would be both unique and meaningful. If a correlation coefficient cannot be computed a period (.) is displayed at the appropriate point in the matrix.

The keyword **ALL** may be used, but this would be appropriate only if **all** the variables were numerical (i.e. if the data included no codes).

The following command, specifying a number of variables created from the gender attitudes data by the **COMPUTE** commands given on pp. 83–5:

```
SPSS/PC: correlation sexism extra neur problems.
```

produced the following display:

```
Correlations:   SEXISM     EXTRA      NEUR      PROBLEM

     SEXISM     1.0000     .0099     .0607     -.1913
     EXTRA       .0099    1.0000    -.2197     -.1151
     NEUR        .0607    -.2197    1.0000      .5641**
     PROBLEM    -.1913    -.1151     .5641**   1.0000

N of cases:     47             1-tailed Signif:  * - .01   ** - .001
```

OPTIONS for the CORRELATION procedure

OPTION 1 requests that missing values be **included**.

OPTION 2 specifies **pairwise deletion**. By default, any case with a missing value for **any** variable contained in a specified matrix is deleted from **all** the correlations in the matrix.

Thus, normally, if case 23 has a missing value for the variable *var006*, the following command:

```
SPSS/PC: correlation var001 to var008.
```

would exclude data for case 23 from all of the 64 correlations in the matrix (whether or not the particular correlation involved *var006*). Thus any cases with missing data for any of the specified variables will be excluded from all of the correlations in the matrix. The above display indicates that the analyses are based on the 47 cases (of the original 50) with valid data for each of the 4 specified variables. But if **pairwise deletion** were requested (**OPTION 2**), a case would be excluded only from those calculations for which it had missing data.

OPTION 3 specifies that the significance of **two-tailed** probabilities (rather than one-tailed) should be indicated by asterisks (* for 1 per cent probability and ** for 0.01 per cent probability).

OPTION 4 requests that the output matrices of the correlations be written to a **results file** (as well as displayed on the screen and printed). Such results matrices can later be used as input for other procedures. Details on the use of this option are given on pp. 112–113.

OPTION 5 requests that, rather than simply displaying the **significance** of the correlation coefficient (with one or two asterisks, where ★ indicates a probability of less than 0.01 and ★★ indicates a probability of less than 0.001), the **exact probability** be displayed (e.g. **probability = 0.0037**). This option also causes the **count** (i.e. the number of cases used in the calculation) to be displayed for **each** correlation coefficient, and this aspect is particularly useful when **OPTION 2** (pairwise deletion) has also been requested.

STATISTICS for the CORRELATION procedure

STATISTIC 1 requests a display and printout of the mean, the standard deviation and the count for each variable.

STATISTIC 2 requests cross-product deviations and covariance (see a statistics textbook for information on these measures).

N.B. Only one **OPTIONS** subcommand and one **STATISTICS** subcommand can be associated with any one **CORRELATION** command (although more than one option or statistic can be specified on each). Thus:

```
SPSS/PC: Correlation var001 to var006
       : /options= 2 5
       : /statistics= 1 2.
```

Writing results (matrix) files

The output from the **CORRELATION** procedure can be written as a **results matrix** on a file which can later be used as **input** to another procedure (such as **REGRESSION**). Using such pre-processed data instead of the original data can save considerable processing time.

OPTION 4 allows a results file to be written from the results of the **CORRELATION** procedure. Such results must form a square matrix (i.e. every variable correlated with every other variable in the variables list). Thus subcommands which include the keyword **WITH** are ignored when matrices are written.

The correlation matrices are written using a standard format. Each coefficient occupies 10 columns (with seven digits to the right of an implied decimal point). Thus each 80-column record (a **record** is a line of data) can include only eight values (although the rows of a matrix can extend over several records). This format will be expected by those procedures which may later use the matrix file as input.

Following the matrix of correlation coefficients, the results file will specify the number of cases on which the correlation coefficients are based. With listwise deletion (the default) a single value for *n* is given. With pairwise deletion, however, a separate *n* is given for each correlation coefficient. This is provided in a separate *n*-matrix which immediately follows the correlation matrix. All values of *n* are included in a 10-column format (this time with all digits to the **left** of an implied decimal point).

The name of the file to which the matrix is written is either "SPSS.PRC" (the default) or a name which has previously been provided on a **Set Results** subcommand. When naming a matrix results file the extension ".mat" is commonly used. Thus if it is intended to write a results file, a suitable name should be devised earlier in the session (before the **COR-RELATION** procedure is started) and a **SET** command issued, for example

```
SPSS/PC: set results= "GENDER.MAT".
```

The following command file will provide a name for the results file before including a data definition file, calculating a series of correlation coefficients and writing the matrix of results in a file:

```
set results= "gender.mat".
include "gender.def".
correlation var001 to var010
/options= 4.
```

This would produce (on a stored file named "gender.mat") a matrix of correlation coefficients with a single *n* value.

10.2 The PLOT procedure

The **PLOT** procedure produces a graphical display of variable frequencies – the frequencies of two variables are plotted simultaneously in a two-dimensional space. Although the procedure is very flexible, it can become rather complex, and the present discussion will therefore be confined to an initial exploration.

Plots and scattergrams

Let us consider two variables *var001* and *var002*. Any one of the respondents in a study will have a value (either a valid value or a missing value) for both variables, and **both** of these values can be displayed simultanously as a single point on a **bivariate** (two-variable) graph.

Suppose a single case has been given the value 3 for *var001* and the value 4 for *var002*. These values can be represented as a single point (**X**) on the following graph.

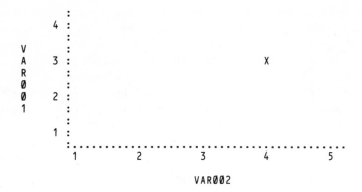

VAR002

Now if we use the **PLOT** command to plot the data points for two variables (*var004* and *var007*) for the 50 subjects from the gender attitudes survey, we obtain the following:

PLOT OF VAR004 WITH VAR007

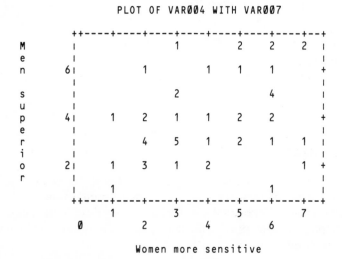

Women more sensitive

Here a 1 indicates a single data point, 2 indicates two data points etc.

This type of graph is often referred to as a **scatterplot**, and we can see that the scatterplot is rather similar to a cross-tabulation table (in this case with 7×7=49 cells).

There is also a noticeable tendency for subjects to have either high values for both *var004* and *var007*, or low values for both of these variables. This suggests that there is a positive correlation between the two variables. This is confirmed by the correlation coefficient (**0.4098**, significant at the **0.01 level**). Those who agree that men are superior in some activities also tend to agree that women are emotionally more sensitive than men.

Using the PLOT procedure

Within the procedure, the word **PLOT** has two separate functions. It acts both as the **command** word and also as an essential **subcommand**. Thus the word **PLOT** must occur twice (at least) for the procedure to work. The **PLOT command** may also include a number of other optional subcommands which affect the format of the graph and the labels and symbols used. Following any optional subcommands, the **PLOT subcommand** contains the list of the variables which are to be plotted in pairs.

The scatterplot of the gender attitude data, above, was produced using the simplest form of the **PLOT** command:

```
SPSS/PC: PLOT plot var004 with var007.
```

The above command contains none of the possible optional subcommands and therefore accepts all of the default characteristics. By default, a single two-dimensional scatterplot of the two variables is produced. The variable named first on the plot subcommand is displayed as the vertical axis. The output display is formatted for the current page size, and the axes are labelled with any variable labels which have previously been specified. Cases with missing data are excluded from scatterplots in which data from one or both variables is missing, but are included in any other scatterplots requested on the plot subcommand.

The optional subcommands

Default parameters can be overridden by using a variety of optional subcommands (there is no **OPTIONS** subcommand for the **PLOT** procedure). Before examining some of the subcommands in detail, let us consider an illustrative example of how various subcommands appear within a single **PLOT** command.

In this case we are going produce a simple scatterplot of the computed variable *sexism* against one of its component variables, *var004* ('**Men superior in many activities**').

```
SPSS/PC: PLOT title= "sexism with 'Men Superior'"
       : /vertical= "sexism" reference (24)
       : /horizontal= "MEN superior" reference (4)
       : /plot sexism with var004.
```

We will first examine this command, statement by statement.

Title is the title given to the graph and is printed at the head.

Vertical in this case specifies the vertical **label** (i.e. the label, sexism, to be given to the vertical axis of the graph) and a **reference** point (here the *sexism* value 24). This will cause a line to be printed on the graph at this point. Up to 10 such reference points can be specified for each of the vertical and horizontal axes.

Horizontal in this case specifies both a horizontal **label** (i.e. the label, Men superior, given to the horizontal axis of the graph) and also a **reference** point, the value 4.

Plot (subcommand) in this case requests a single bivariate (i.e. two-variable) plot of *sexism* with *var004*. Lists of variables can be included in the plot subcommand in the usual way.

The above **PLOT** command produced the following output:

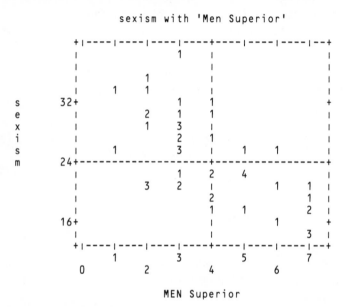

Notice that the scale of the axes has been adjusted automatically to conform to the page size.

The specified title is included as a heading (the default title in this case would have been **'plot of sexism with var004'**).

Reference lines have been drawn, at the points requested, on the horizontal and vertical axes.

The output from this plot would seem to indicate that there is a strong association between responses to the **'men superior'** item on the gender attitudes questionnaire (coded as *var004*) and the composite *sexism* score. A **CORRELATION** analysis of these variables revealed a highly significant correlation (-0.6558), indicating that a high *sexism* score is associated with a **low** score for *var004*. This indicates that 'sexist' respondents are more likely than 'non-sexist' respondents to agree with the statement that **'there are many activities in which men are clearly superior to women'**. However, it must be remembered that the **COMPUTE** command used to create the score for *sexism* included *var004* as one of the elements.

The Missing subcommand

Normally, missing data are not included in the plot. Deletion is plotwise (equivalent to pairwise). But listwise deletion can be specified, using:

```
/Missing= listwise
```

and missing data can be **included**, using:

```
/Missing= include
```

The Format subcommand

The **Format** subcommand specifies the type of plot produced. The default produces one or more **bivariate scatterplots**.

If **overlay** is specified then all variables specified on the next **PLOT** subcommand are plotted **together** on a single graph.

```
/Format= overlay
```

For an overlay plot, **PLOT** automatically selects a symbol for each layer and also selects symbols to represent superimposed points. A key is provided on the display.

Other formats available include **contour** plots and **regression** plots (see *SPSS/PC+ Manual* pp. C114–C119).

The Symbols subcommand

The **Symbols** subcommand is used to specify which symbols should be displayed in the body of the graph. The default is alphanumeric. Thus 1 represents one data point, 2 represents two data points and so on. After 9 the symbols change to upper-case letters A to Z. All values of 36 or above are represented by an asterisk ($\star$). These symbols can be changed by the user. Thus, in the following subcommand:

```
/symbols= "*"
```

the user is specifying that any number of data points (one or more) occurring at a particular point in the graph space should be marked with an asterisk. Here:

```
/symbols= "*,@,?,&"
```

the user is specifying that an asterisk ($\star$) should be used if there is one data point present, that @ should be printed if two data points are present, etc. Up to 36 separate symbols can be specified, and the printed plot will include a key.

The Vsize and Hsize subcommands

The **Vsize** and **Hsize** subcommands specify the vertical and horizontal sizes of the display frame used for the plot. Default values depend on the current page size of the paper used (either the SPSS/PC+ default or a size which has been set by a **SET** command). For normal 80-column printed pages the default values are **Vsize** = 16 (rows); **Hsize** = 38 (columns). Changes can be made in following way:

```
/Vsize= 20   /Hsize= 45
```

The Title subcommand
The **Title** subcommand simply provides a title for a graph. Thus:

```
/Title= "Sexism with 'Men Superior'".
```

The title can be up to 60 characters long.

The Vertical and Horizontal subcommands
These subcommands can include a title for the axis. Thus:

```
/Vertical= "Sexism"
```

and they may also specify the minimum and maximum values to be plotted (default plots **all** values); any data which fall below a specified minimum or above a specified maximum are simply not plotted. Thus:

```
/Horizontal= "Men superior " MIN (2) MAX (6)
```

These subcommands can also be used to specify values at which **reference lines** will be printed, e.g.

```
/Horizontal= "Sexism"   reference (24)
```

For overlay plots, when two or more variables with different ranges are to be plotted on the same axis, it is often useful to standardize scores so that each variable is effectively reduced to the same scale. This is done on the same subcommand, by specifying **standardize**. Thus:

```
/Horizontal= standardize
```

Finally, it is possible to specify that all plots on a PLOT subcommand list be plotted on axes with identical scales (normally the scales would differ, with each tailored to the particular data being plotted). This is achieved by specifying uniform. Thus:

```
/Horizontal= uniform
```

It can be seen that rather a lot of information could be contained on a single /**Horizontal** or /**Vertical** subcommand. Thus:

```
/Vertical= "Salary"   MIN (6000) MAX (10000) standardize
```

N.B. The subcommands **Missing, Vsize, Hsize** and **Symbols** apply to **all** plots within a **PLOT command** and can therefore be specified only **once** in any **PLOT** command.

If you find all these subcommands (and there are others) intimidating, take heart. For most purposes the defaults will probably serve you well. Remember that we produced an interesting scatterplot with just:

```
SPSS/PC: PLOT PLOT= var004 with var007.
```

Multiple plots can be specified by a single command. For example,

```
SPSS/PC: PLOT PLOT= var001 with var002;
       : var003 var004 to var 006 with var010;
       : var008 with var009.
```

(Notice that a semi-colon, not a slash, is used.) The above command requests six bivariate plots.

Although, as in this example, several **Plot sub**commands can be included in a single **PLOT command**, some subcommands (such as **Title, Horizontal, Vertical** and **Format**) affect only the **Plot** subcommand which immediately follows them. Therefore, even if default characteristics are to be changed in exactly the same way for successive **Plot** subcommands, the appropriate changes must be specified **each** time.

Thus, to request different titles, but the same reference lines, for two successive **Plot** subcommands, a command of the following kind would be needed:

```
SPSS/PC: PLOT  / title= "First two variables"
       : /horizontal= reference (4)
       : /vertical= reference (3)
       : /plot= var001 with var002
       : /title= "Second two variables"
       : /horizontal= reference (4)
       : /vertical= reference (3)
       : /plot = var003 with var004.
```

This one **PLOT** command statement contains two separate **Plot** subcommands (and six additional subcommands). The final subcommand must always be a **Plot** subcommand.

A special pairing can be requested (similar to the special pairing we met in **T-TEST OPTION 5**), by specifying **(pair)** at the end of a list containing the same number of variables on either side of the **WITH** keyword. Thus, for example,

```
SPSS/PC: PLOT plot age with var001 to var006 ;
       : var003 to var006 with var007 to var010 (PAIR);
       : var008 with var009 var010.
```

Here *var003* would be paired **only** with *var007*, *var004* would be paired only with *var008*, and so on. The complete command would therefore produce 12 plots.

No statistics are available for the **PLOT** procedure.

11   *Log Files and Listing Files*

In this chapter we will introduce two new types of file, **log files** and **listing files**, and explore some further features of the **REVIEW** editor.

The material in this chapter will bring the number of different types of file so far discussed to six (a brief guide to the features of the file types used in SPSS/PC+ is given in Appendix H).

11.1 Types of file

Data files contain only raw data.

Command files contain **commands**, including operation commands, data manipulation and transformation commands, and procedure commands (for statistical analyses, etc.). One special type of command file, the **data definition file**, contains the raw data (or an instruction on where to find the data), together with a data dictionary which includes information about the location and format of the data, variable labels and value labels, and missing values. Other command files may include a series of procedure commands for the analysis of data by batch processing, or a series of operation commands which change certain operating parameters of the SPSS/PC+ system.

Results files are written from the output of various procedures (including **CORRELATION**) when a particular **OPTION** is specified. Results files are rectangular matrices of means, standard deviations, correlations and the like, which may be used as input to other statistical procedures. Results files are written to a file named "SPSS.PRC" unless an alternative name has been specified by using an appropriate **SET** command. Results files are stored in ASCII (character) format and so may be edited with **REVIEW**. They can also be printed from **DOS** using, for example,

```
C>   type gender.mat>prn
```

System files contain data, the data dictionary, and certain system variables. However, the information is stored in a special binary format. This means that system files are efficient to use, but system files **cannot** be directly edited.

Thus **data files** and **command files** are written by the user and can be edited. **Results files** are created by the system when the user issues an appropriate request. They, too, can be edited. **System files**, however, are written by the system following a **SAVE** command, and **cannot** be edited.

This chapter will introduce two new types of file. SPSS/PC+ automatically maintains two separate files of the current SPSS/PC+ session. The **log file** provides a record of the commands issued during the session. The **listing file** contains a copy of the output which has been produced during the session.

Log files consist of a log (akin to a ship's log) of the **commands entered** during the current interactive session with SPSS/PC+, together with information that a warning or error was issued, and references to output page numbers. The log is maintained on a file named (as default) "SPSS.LOG". When a session begins, the existing log file is initialized (and the contents therefore erased). Any command entered during the session is recorded as the session proceeds, and this provides an editable record which can be useful in monitoring progress and diagnosing errors. The log can also be made the basis of a new command file, and is therefore useful when we wish to repeat procedures (perhaps with the contents edited in such a way that a procedure is run with different variables, or perhaps with different **OPTIONS**). Rather than laboriously retyping the previously used commands, we could instead **REVIEW** the log file and edit the sequence of commands which it contains. This edited material can then be written as a command file which would be **INCLUDED** so that further batch processing of the current active file would take place.

Listing files consist of a copy of the **display output** for the current session, together with any warnings or error messages which have been issued. The pages within the listing file are numbered with the same numbers that appear on any printed copy of the output. The listing is available on a file named (by default) "SPSS.LIS". A listing file can be edited, perhaps by deleting certain results or by annotating the text, so that the contents can be printed in a form specified by the user.

For both the log file and the listing file, a name other than the default name can be specified at the beginning of the session, using the **SET** command, for example

```
SPSS/PC: set listing "gender.lis".
```

The use of such **SET** commands prevents the loss of the records at the beginning of the **next** session when any files with the default names "SPSS.LIS" and "SPSS.LOG" are initialized. If alternative names have been provided, the log and listing files created during the session will remain on the disk until deliberately deleted or overwritten by another file which has been given the same name. Thus to keep a permanent record of the commands and output from each session, **different** names must be given to the log file, and to the listing file, on each occasion.

If the log or listing name is **SET during** a session then the named files will contain the log or listing **from that time on**. Records of the earlier part of the session will be retained in the separate files with the original name(s).

Listing files and log files are stored in ASCII form and can be displayed and edited. The contents of these files can also be printed (using **DOS**) and read into word-processor files.

In this chapter we will explore the ways in which **log files** and **listing files** can be edited using **REVIEW**. For handling the log files, in particular, **REVIEW** is the editor of choice, allowing editing to take place without bringing the current SPSS/PC+ session to an end. General information about using other editors, such as the **DOS** editor **EDLIN** or word-processing systems, is provided in Appendix B.

11.2 Entering files into REVIEW

To edit a **log file** or a **listing file**, **REVIEW** can be entered at any time during an interactive SPSS/PC+ session. This will not disturb the active file in any way. After editing with **REVIEW**, the user can return to SPSS/PC+ to continue the session.

To enter **REVIEW** you may simply issue the command:

```
SPSS/PC: review.
```

This will produce a screen display split into a upper half and a lower half (these are sometimes referred to as the upper **window** and the lower **window**). The upper window will show the end of the most recent display produced by any procedures used in this session (i.e. the most recent part of the listing file). The lower window will show the most recent commands which have been issued and other information including references to errors and to output page numbers (i.e. the most recent part of the log file).

Initially, the cursor will be in the first column of the bottom line of the lower (log) window. It can be moved around within this window for editing purposes. As the cursor is moved horizontally a number in the lower right-hand corner of the screen changes. At any time, this identifies the current cursor **column** position. The cursor will not move horizontally beyond the right-most character of any line, but additional characters or spaces may be added to an existing line.

Moving the cursor upwards will have the effect, when the top of the window has been reached, of scrolling to an earlier part of the file. In this way access is provided to parts of the file which lie beyond the initial window display.

Changing windows

The cursor can be made to switch between the lower log window and the upper listing window by pressing <F2>. One key press will remove the cursor from its initial position within the log window and move it into the upper (listing) window. It can be moved within this window to edit the listing file. Moving the cursor to the top of the window will cause scrolling to an earlier part of the listing file.

Changing windows causes the file name message at the bottom of the screen to change from "SPSS.LOG" to "SPSS.LIS", (or to the name previously specified for the listing file). Pressing <F2> again will cause the cursor to move back into the log window and the name of the log file will return.

Changing the window size

The split screen arrangement allows the user to directly compare the commands issued with the output which resulted from the command. This top and bottom split, however, limits the display size available for each of the files. Initially, the top (listing) window contains only 12 lines and the bottom (log) window 10 lines. These window sizes can be changed.

You may remember that it is possible to get a display of the various **REVIEW** functions simply by pressing <F1>. Do this and you will see the full list of the **REVIEW** function commands available. Since many of these are used only rarely, and help from the display can be obtained at any time during an edit, it is not worth trying to learn all of the various commands. The immediate concern at this stage is to change the size of the display window. It can be seen from the **REVIEW** command table that to **Change Window Size** we need to press <^F2>. When this is done the message

```
Number of lines for upper window:
```

will appear, and the cursor will be seen flashing inside a box. The upper window is initially set at 12 lines, but this can be changed to any number between 3 and 19. A high number will increase the size of the listing window and correspondingly reduce the size of the log window. A low number will have the opposite effect.

Entering a single (log or listing) file into REVIEW

If the user needs to examine or edit only **one** of the two files then the split window can be avoided altogether. A specific request can be made to **REVIEW** either the log or the listing file:

```
SPSS/PC: review listing.
```

or

    SPSS/PC: review log.

These commands will produce a full-screen display of the most recent part
of the file which has been requested.

11.3 Paging and searching

Although it is possible to move within the file using just the cursor, there
are more efficient ways of moving around within long log or listing files.
Thus we can search for a particular set of characters (a string – for example
the name of a certain variable) or we can direct the cursor to a particular
SPSS/PC+ page. Paging the listing (or output) file will be especially easy if
we have a copy of the printed output, since the SPSS/PC+ page numbers
on the file will be the same as the page numbers on the printout.

Finding a page

To find a particular numbered SPSS/PC+ page in a listing file, press
< F6> and reply to the message:

    Page number:

with the number of the page you wish to find. The cursor will then be
placed at the beginning of that page.

Searching for strings

Pressing <F6> will initiate a **forward** search through the file (from the
current cursor position) for a specified string of characters. The computer
will present the message **Search string:**. The string name must be entered
in **exactly** the form required. Upper-case and lower-case characters are *not*
equivalent: specifying 'Neur' will *not* find 'neur'. The computer will search
forward through the file for the first occurrence of the specified string and
the cursor will flash on the first character of that string.

To find a second occurrence of the same string, <F6> has to be pressed
again. Once a string has been entered it becomes the default string, so that
any subsequent occurrence of the same string can be found merely by
pressing <F6> (followed by <RETURN>).

< ↑ F6> will produce a **backward** search through the file (i.e. the search
will start from the current cursor position and proceed backwards through
the file).

11.4 Editing log and listing files

REVIEW is a general-purpose editor and thus log files and listing files can be edited in much the same way as other files. Remember that there are two modes; insert (the **Ins** message will appear at the bottom of the screen) and overwrite (or insert off). When insert is **on**, any typed characters are inserted at the cursor position, and existing characters are pushed to the right. When insert is **off**, existing characters to the right of the cursor are overwritten by any newly typed characters.

New blank lines can be inserted after a current line (i.e. the line with the cursor on it) by pressing <F4>. The current line can be deleted by pressing <⌃ F4>.

In addition to these basic editing functions of **REVIEW**, there are a number of more complex functions which have a special relevance to the editing of log files and listing files.

One of the most useful features of **REVIEW**, especially when editing log files, is its ability to replace one sequence of characters (a string) with another.

Changing strings

Strings of characters can be changed. Suppose that we wish to repeat a set of procedures that have already been performed using *var001*, but this time using *var002*. We could obtain a log of the current session, containing the full sequence of commands issued for the analysis of *var001*, and then replace *var001* with *var002* throughout the relevant part of the log. We could then use the **REVIEW** <F7> function (at the beginning and at the end of the material to be included in the new file) to make a block within which each command would now refer to *var002*. This block would then be written to disk as a new command file.

To **change forwards** from the current cursor position, we would issue the command <F5>. In response to the prompt **Old string:** we would enter the string to be changed (i.e. in this case *var001*). After pressing <RETURN> a **New string:** prompt would appear. We would then enter the string to be substituted (in this case *var002*).

There will now be a search for the old string, starting from the position of the cursor before the <F5> command was issued. When the first occurrence of the string is located the user is presented with a screen display listing a number of options as described in table 11.1.

Table 11.1

C=chg&next   A=chg all   N=skp&nxt   S=stop   X=change&stop

The appropriate key – C, A, N, S or X – must now be pressed
C replaces the string and proceeds to next occurrence of old string
A replaces this and every (subsequent) occurrence of old string
N leaves this occurrence intact and proceeds to the next
S leaves this occurrence intact and stops the search
X replaces this occurrence and then stops the search

Following a **C** or **N** response, the same list of options will be presented again when the next occurrence of the old string is found.

In the example quoted (where we wish to change **all** the occurrences of *var001* to *var002*) we could reply to the option display by keying **A** (= change this and all subsequent occurrences). The changes will be made automatically throughout the remainder of the file, and the computer will then issue a message detailing the number of changes made (e.g. **8 changes made**).

If a **change backwards** is required, rather than a **change forwards**, then the replacement procedure should be initiated with < ↑ F5> rather than with <F5>.

Dealing with square bracket lines within log files

It is likely that many of the lines of text within the log file will be preceded by a square bracket [. Such lines may contain records of error messages, information which has been read into SPSS/PC+ via an **INCLUDE** command, or output page numbers. If the log file is being edited in order to create a new command file then special attention should be paid to the lines with a square bracket.

If such a line is to be an **active** element in the new command file the bracket must be deleted. Lines with a bracket will not be executed. On the other hand, lines which refer to previous errors or to page numbers can be safely retained within the file, but such lines must **retain** the initial bracket if errors are not to occur. As long as the bracket is retained at the beginning of each of these lines they will have no effect. They will not disturb the system or produce errors when the file is subsequently made active with an **INCLUDE** command.

Using SPSS/PC+ HELP from within REVIEW

When editing log files, especially, it is often useful to be reminded of the alternative **OPTIONS** and **STATISTICS** which are available for particular procedures. Having analyzed data with one selection of such possibilities the user may wish to explore others. The SPSS/PC+ **HELP** command, which supplies such on-line information, is accessible from within **REVIEW**. Thus when revising the **STATISTICS** subcommand for a previously submitted **FREQUENCIES** procedure, you can request an on-screen reminder of the statistics available. To do this from **REVIEW** you would press $< \uparrow F1>$. When the prompt **SPSS/PC+ help topic?** appears, you issue an appropriate request, e.g.

```
SPSS/PC: help topic? frequencies statistics. <RETURN>
```

The system would respond by displaying a list of the statistics available for the **FREQUENCIES** procedure. Pressing <SPACE> will return the system to **REVIEW**.

Similarly, the user can be reminded of the variables which exist in the current active file. Issuing the **REVIEW** command <∧F1> will produce a display of all the variables in the active file and, again, pressing <SPACE> will return the user to **REVIEW**.

Using the edited log file

When the log file has been edited, the block of revised commands can be written to a new command file (by first pressing <F9> and then either entering a new file name or pressing <RETURN> to accept the default name "REVIEW.TMP"). To exit from **REVIEW** the command <aF10> should be issued. The new file can be brought into use by naming it in an **INCLUDE** command, for example

```
SPSS/PC: include "newvar.sps".
```

There is another way of achieving the same result. Having marked the beginning and end of the block (using <F7>), instead of writing the block as a separate file and then using an **INCLUDE** command, it is possible to exit and run the block in a single step. This is done simply by pressing <F10>. There is a prompt, followed by the default name for the block: **Name for block: REVIEW.TMP**. A specific name may be provided for the block, or <RETURN> may be pressed to accept the default name. When the block has been written to disk, an exit is made from **REVIEW**. Any commands written within the block are executed and the user is then returned to the SPSS/PC: prompt. The block will remain on the disk until deleted or overwritten. If the default name was accepted, then it will be overwritten by the next temporary review file ("REVIEW.TMP").

Unless you name the **edited** version of the log file "SPSS.LOG" (or the name you have previously specified for the log file) all of your editing actions will have no effect on the actual (i.e. the original) log file which is automatically written by the system.

Editing listing files

All or part of the recent output (the results or listing) can be edited and written to a separate file for later printing. All of the normal editing features may be employed. Thus the content and layout of a results listing (a crosstabulation, for example) can be altered with total freedom. Parts of the text or of the output data can be deleted, and extra text material can be added.

The simplest way to print the edited listing file is to exit from SPSS/PC+ and return to **DOS**. Exit from SPSS/PC+ by using the **FINISH** command and request a print of the listing file (or an edited extract which has been written to a named file) with a suitable command, for example

```
C> type spss.lis>prn
```

If the default name ("SPSS.LIS") has been retained then such a print will only be available until SPSS/PC+ is loaded again (at which time the "SPSS.LIS" file is initialized).

The listing file can be used directly by a word-processing program which is resident on the current machine, or it can be copied to a diskette placed in **drive A** (for example, to be taken to another compatible machine for word-processing). To copy the existing listing file, the following command is used:

```
C> copy spss.lis a:listing
```

DOS will now copy the listing file ("SPSS.LIS") from the current **DOS** directory to a file named "listing" on **drive A**.

11.5 Some final points on REVIEW

From within **REVIEW**, the user can ask for a file on disk to be directly inserted for inspection and editing. This is done by pressing <F3> and replying to the prompt **File to insert:** with the name of the relevant file. Certain files, of course (like system files), are not editable.

At any point during a **REVIEW** session, the user can request a list of the files in the current directory. This is done by pressing <aF1>. The computer replies with a prompt **File specification:**.

There are now three options:

- If <RETURN> is pressed, the computer will display the names of **all** of the files in the current **DOS** directory. The list will appear within a window on the screen, and only a small section of the directory will be displayed at any one time. The list will proceed from the oldest files to the most recent, and a number of key presses may need to be issued in order to obtain a display of the end of the directory (where the most recently created files are listed).

- The name of one particular file can be supplied. Information on that file (only) will then be displayed in the window.
- A request can be made for a display list of a particular group of files. An asterisk ($\star$) acts as a wildcard. Thus specifying "**$\star$.sps**" would produce a list of every file in the current directory with the file extension **sps**. Specifying "**gender.$\star$**" would produce a list of all **gender**, files whatever their file extension. A **?** symbol acts as a wild character. Thus "**gend????.$\star$**" would produce a list of any file with **gend** as the first four letters (with or without any other characters), and with any file extension. The files "**gender.sps**" and "**gendatt.mat**" would be included in the list requested by this specification. If, with any wild characters, the length of the file name (without extension) is less than eight characters, then only files with names up to and including the specified length will appear in the list. Thus "**gend??.$\star$**" would produce a list which includes "**gender.sps**" but not "**gendatt.mat**".

In Table 4.1 a list was presented of the **REVIEW** commands which had been encountered up to that stage. In this chapter we have met several more. Table 11.2 provides a summary of all of the **REVIEW** commands which have been met.

In this chapter we have considered the use of two types of file which are automatically written by SPSS/PC+ during each session as a record of events (both input, the log file and output, the listing file).

Table 11.2 REVIEW commands

Command	Key(s)
Information	
REVIEW help	<F1>
SPSS/PC+ HELP from within REVIEW	< ↑ F1>
Information on variables in current active file	<^ F1>
List files	<aF1>
Change display	
Switch between upper and lower window	<F2>
Change window size	<^ F2>
Delete and insert	
Insert file	<F3>
Insert a blank line after current line	<F4>
Delete current line	<^ F4>
Find and change	
Find a numbered output page	<^ F6>
Forward search for string	<F6>
Backward search for string	< ↑ F6>
Change string forwards	<F5>
Change string backwards	< ↑ F5>
Block, write and exit	
Mark block (beginning *and* end)	<F7>
Write block to disk	<F9>
Write file to disk	< ↑ F9>
Exit from REVIEW	<aF10>
Exit and run the block	<F10>

12 Statistical Analysis IV: ONEWAY and ANOVA

Both of the procedures **ONEWAY** and **ANOVA** perform analysis of variance (anova). This statistical test compares the values of a dependent variable for cases which fall into a number of different groups. The independent (between subjects) t-test procedure does the same thing, but only two groups can be compared in each t-test analysis. A one-way analysis of variance can compare two, three or more groups classified in terms of a single independent variable. A two-way analysis of variance compares groups formed by classifying cases simultaneously on two independent variables. If the SPSS/PC+ analysis of variance is to include just one independent variable a one-way anova is performed using either **ONEWAY** or **ANOVA**. With two independent variables a two-way anova is performed; and so on for three-way anova, four-way anova, etc. For an anova with two or more (and up to 10) independent variables, the procedure **ANOVA** must be used.

As well as being a method for the statistical analysis of data, anova refers to an approach to experimental design. Many studies are devised to conform to a particular type of anova, so that there is a close association between the nature of the study and the analysis which is eventually performed on the data obtained. The procedures **ONEWAY** and **ANOVA** are appropriate for designs in which **different cases** are found within each of the conditions. This is therefore similar to the **independent** t-test. There are anova designs in which comparisons can be made between different variables for the **same set of subjects** (as in the **paired** t-test). Such repeated measures designs cannot be analyzed by the procedures detailed in this chapter, however. If you wish to analyze data from a study which follows such a design you should use the **MANOVA** procedure which is included within the *SPSS/PC+ Advanced Statistics* optional enhancement.

ONEWAY and **ANOVA** perform an analysis of variance to assess how much of the variation (variance) of a dependent variable can be assigned to different sources, including the independent variable(s) and any inter-actions between such variables. For example, it might be found that a measure of psychological difficulties, *problem*, (the dependent variable), varies systematically with two independent variables, *sex* and *faculty*. These variables would therefore be seen as significant factors in determining the scores of the dependent variable. It might also be found that there is a significant **sex×faculty** interaction such that, for example, high *problem* scores tend to be found for male science students but for female arts students.

Anova calculates the variance of the scores within conditions (for example, within **each** of the three faculties) and between conditions (in this example, between the three different faculties). If the **within**-conditions variance is low compared to the **between**-conditions variance then it would seem that the scores for the dependent variable tend to be different in the different conditions. This would suggest that the independent variable is able to explain a significant part of the overall variation between the scores.

Anova partitions the total variance into the treatment variance (i.e. that which is explained by the independent variables and any interactions between them), and the residual variance (i.e. that which is **not** explained) and compares these two values. If a substantial part of the total variance relates to differences between the conditions of an independent variable then that variable would seem to explain a significant part of the variance. The effect of this variable would therefore be seen as significant (i.e. it would have a low 'chance probability').

Taking the simplest example, the one-way anova, the stages in the computation can be outlined in the following way. First the between-groups and within-groups sums of squared deviations from the mean are calculated. Each of these values is then divided by the relevant degrees of freedom to yield a mean square, and when the between-groups means square is divided by the within-groups mean square, a variance ratio (**F-score**) is obtained. If this value is sufficiently high (i.e. of low probability) then it will be statistically significant, and the effect of the relevant independent variable will have been shown to be significant.

12.1 The ONEWAY procedure

A one-way anova deals with only one independent variable. The variance **between** the groups formed by this independent variable is compared with the residual variance, i.e. the variance **within** the groups.

The simplest form of the **ONEWAY** command is:

```
SPSS/PC: ONEWAY var003 BY fac(1,3).
```

Here *var003* is the dependent variable. The numerical values of *var003* will be compared for different groups of cases (subjects). There is no requirement that these groups be of equal size. The groups will be formed from the categories corresponding to particular values of *fac* between 1, which is specified in the command as the **minimum** value to be used in forming groups, and 3, which is specified as the **maximum** value to be used. The minimum and maximum values must be integers (i.e. whole numbers, without a decimal point).

Because of the nature of the statistical test to be performed, only one independent variable may be specified in a single **ONEWAY** subcommand (if you wish to perform a similar analysis with more than one independent variable then the alternative procedure **ANOVA** should be used).

Although only one **independent** variable can be specified on a **ONEWAY** subcommand, up to 100 **dependent** variables may be specified. The normal variable listing conventions are permitted. Thus:

```
SPSS/PC: ONEWAY var001 to var010 BY fac(1,3).
```

This would produce 10 one-way analyses of variance. Each of the 10 dependent variables would be compared across the three *fac* groups. An analysis of variance will tell us whether there is a significant variation in the dependent variable (say, *var001*) between the three *fac* groups analyzed.

To produce the following example, a new variable *sextract* was computed from the gender attitudes survey data. The **COMPUTE** command was the same as that for *sexism* (p. 85) except that it did not include *var003* as an element. The raw scores for *sextract* were then recoded to form three groups, 'low *sextract*', 'medium *sextract*' and 'high *sextract*':

```
SPSS/PC: recode sextract (10 thru 18=1) (19 thru 23=2)
       : (24 thru 35=3).
```

The **ONEWAY** procedure was then used to see whether the three *sextract* groups differed significantly on the variable which had **not** been used in the computation of the original *sextract* score:

```
SPSS/PC: oneway var003 by sextract(1,3).
```

This produced the following output:

```
- - - - - - - - - - O N E W A Y - - - - - - - - - -

    Variable   VAR003     Women use sexual dis. as excuse
    By Variable  SEXTRACT   Recode 10 to 18 - 19 to 23 - 24 to 35

                        Analysis of Variance

                          Sum of      Mean        F      F
           Source    D.F.  Squares    Squares    Ratio  Prob.

Between Groups     2     20.0761    10.0380    4.1124  .0229

Within Groups     45    109.8406    2.4409

Total             47    129.9167
```

Because there are 48 valid cases (N) the value of the total degrees of freedom is ($N - 1$) 47. Because there are three groups (k) the between groups degrees of freedom is 2 (i.e. $k - 1$). The difference between these values ($47 - 2 = 45$) gives the degrees of freedom assigned to the residual (**within groups**) variance. The sums of squares have been calculated for the whole population (total sum of squares) and for the between-groups and within-groups sources. Dividing the between-groups and within-groups sums of squares by their respective degrees of freedom has yielded a mean square value for each. Dividing the between-groups mean square by the within-groups mean square has produced an F-ratio of 4.1124, corresponding to a probability of **0.0229**. There is thus a significant effect.

With no **OPTIONS** or **STATISTICS** subcommands in effect, the default table is produced, as above, for each anova requested. The table contains the F-value and its probability. It also includes the degrees of freedom and both the sum of squares and mean squares for each of the two sources within-groups and between-groups. Although the significance of the overall variation between groups is given by the F-statistic, no tests to determine differences between individual pairs of groups have been performed. Cases with missing values on either the independent or dependent variable are excluded (pairwise).

A number of subcommands may be included within the **ONEWAY** command. **RANGES, STATISTICS** and **OPTIONS** will be described. Two of the subcommands available for **ONEWAY** (**POLYNOMIAL** and **CONTRASTS**) will not be discussed here (full specification is given in the *SPSS/PC+ Manual*, pp. C107–C113).

The Ranges subcommand

When an independent variable (such as *sexism*, recoded into three categories) is shown to have a significant effect, we know that there is an overall difference in the values of the dependent variable across the groups. We do not know, however, which **pairs** of groups differ significantly. This can be assessed using a range test which can be requested using the **Ranges** subcommand.

Ranges specifies one of seven available tests to be used in identifying which pairs of groups (within the range specified for the independent variable) have significantly different means.

If three groups are specified for the independent variable then **three** comparisons will be made. Comparisons will be made between:

Group 1 and Group 2   Group 1 and Group 3   Group 2 and Group 3

- 4 groups would produce 6 comparisons
- 5 groups would produce 10 comparisons

- 6 groups would produce 15 comparisons
  etc.

The range tests available include the **Duncan multiple range test (Duncan)**, the **least significant difference (LSD)** and the **Student–Newman–Keuls (SNK)**.

The default alpha (significance level) for all range tests is 0.05, but for certain tests (including **Duncan** and **LSD**) this level may be changed.

Full details of the range tests available can be obtained from the *SPSS/ PC+ Manual* (pp. C110–C113). On-screen help can be obtained by the command:

```
SPSS/PC: help oneway ranges.
```

The **Ranges** subcommand is used in the following way:

```
: /ranges=duncan
: /ranges=lsd(.01).
```

Each range test has to be specified, as above, on a separate **Ranges** subcommand. For the first range test specified (**Duncan**), the default alpha level (0.05) is accepted. For the second test (**LSD**), however, the value is changed to 0.01.

In the printout for a range test, the means of the groups are listed in ascending order and a grid is produced, with asterisks used to show which means are significantly different. Thus, in the analysis of *var003* by *sextract*, the following command:

```
SPSS/PC: oneway var003 by sextract(1,3)
       : /ranges=duncan.
```

produced a **Ranges** output which included the following table:

```
(*) Denotes pairs of groups significantly different at the .050 level

    Variable   VAR003      Women use sexual dis. as excuse

                           G G G
                           r r r
                           p p p
    Mean       Group       3 1 2

    3.6000     Grp 3
    4.8421     Grp 1       *
    5.1429     Grp 2       *
```

This output indicates that the mean values of *var003* for Groups 1 and 3, and 2 and 3, are significantly different but that there is no significant difference between the mean values of this variable for Groups 2 and 3 on the range test used.

Thus the 'high *sextract*' group differs significantly from both the 'medium *sextract*' groups and the 'low *sextract*' group (only) in the degree of agreement with the statement '**Many women use sex discrimination as an excuse for their own shortcomings**'. From the Group means we can see that the 'high *sextract*' group (Group 3) have a **lower** mean, i.e. they tend to agree more with the statement.

OPTIONS

A number of options are available for the procedure **ONEWAY** (for a full list use **help oneway options.**).

Some **OPTIONS** affect the way in which missing values are treated. **OPTION 1** specifies that missing values should be included in the analysis, and **OPTION 2** specifies that missing values should be excluded listwise. **OPTION 3** requests that variable labels should be suppressed, and **OPTION 6** specifies that value labels should be used to label the groups. Three of the options deal with the reading and writing of matrices.

Matrices

In Chapter 10 it was shown how output from the **CORRELATION** procedure could be written as a results file (by specifying the **CORRELATION OPTION 4**). The **ONEWAY** procedure can also produce a results file, and it can also use a results file produced in a previous analysis.

Writing results files

To request that an output matrix from the **ONEWAY** procedure be written to a separate results file, **OPTION 4** is used. The default name for such a file is "SPSS.PRC", but a **SET RESULTS** command can be used to specify an alternative name (the extension **.mat** is often used). The following command sequence (using *sexism*, recoded in the following way – 14 through 20 = 1; 21 through 26 = 2; 27 through 38 = 3) was used to execute a **ONEWAY** procedure and write a results file:

```
SPSS/PC: SET RESULTS= "GENDER.MAT".
SPSS/PC: oneway var007 BY sexism(1,3)
       : /OPTIONS= 4.
```

The **SET RESULTS** command arranges for the results of the procedure which follows to be written to a results file named "gender.mat". The **ONEWAY** command requests an analysis of variance of the values of the dependent variable (*var007*) for three groups defined by the specified values of the independent variable *sexism*. The **OPTIONS** subcommand requests that the results matrix be written to the file named on the **SET RESULTS** command.

The above command produced the following "gender.mat" file:

```
14.00          17.00          17.00
 5.0000         4.0000         3.5294
 1.5689         1.8371         1.8068
```

This results matrix contains vectors, for each group, of:

Line 1   The **count**
Line 2   The **mean**
Line 3   The **standard deviation**

All vectors are written over 10 columns, with two places to the right of the decimal point for the **counts**, and four places to the right of the decimal point for the **means** and **standard deviations**.

Because 10 columns are occupied by each vector, the maximum number of vector values (one per group) which can be recorded on a single line (a record) is 8. If more groups are involved in the analysis then the matrix will include two or more records for each type of vector.

Reading matrices

When a results matrix has been stored as a results file it can be used to provide the input for certain procedures. A suitable results file can be read into **ONEWAY** by specifying the name of the file and including the relevant **OPTIONS** subcommand (in this case, **OPTION 7**):

```
SPSS/PC: data list file= "gender.mat" matrix/ var007 sexism.
SPSS/PC: oneway  var007 by sexism(1,3)
       : /options= 7.
```

In this example, the **DATA LIST** specifies the name of the file from which the input is to be taken. It also informs the system that this is a matrix file and supplies names for both the dependent variable and the independent variable to which the matrix information relates. The **OPTIONS** command informs the system of the type of information included in the matrix.

If these commands followed the **ONEWAY** procedure described above, they would cause the system to read the results file and repeat the original analysis, but this time using the matrix information rather than the original data. New names could be given to the variables read from the file. The results produced would, of course, be the same as those previously obtained.

In the **ONEWAY** procedure, two **OPTIONS** are available for reading matrices, and these lead the computer to expect the matrix to contain information in one of two arrangements.

OPTION 7 leads the computer to expect a matrix in the same format as that which can be output from **ONEWAY OPTION 4** (i.e. counts, means and standard deviations in the 10-column format specified above).

OPTION 8, however, leads the computer to expect **four** vectors: **counts** (for each group), **means** (for each group), **pooled variance** (within-groups mean square: one value only), and **pooled variance degrees of freedom** (one value only).

STATISTICS

Additional statistics can be obtained by using the **STATISTICS** subcommand. Three sets of additional statistics are available.

STATISTIC 1 requests further descriptive statistics for each group, including the minimum, maximum, mean, standard deviation and standard error. For details on the other available statistics, get on-screen information with **help oneway statistics.**

12.2 The ANOVA procedure

The **ANOVA** procedure is used to obtain analyses of variance for a numerical dependent variable on up to 10 categorical (i.e. classification) factors (i.e. up to 10-way analysis of variance can be performed). There must be different subjects in each of the conditions. Analyses of variance can take several forms (factorial, hierarchical, regression), each of which can be specified by a suitable **OPTIONS** subcommand.

In this introduction only the simpler uses of the procedure will be discussed. Those who are unfamiliar with analysis of variance should consult a statistical text.

The simplest form of the **ANOVA** command is:

```
SPSS/PC: ANOVA var007 BY sexism(1,3).
```

This would produce a one-way analysis of variance, with the recoded *sexism* as the independent variable and *var007* as the dependent variable. *Three* independent groups are specified (those taking the *sexism* values 1, 2 and 3).

In the following **ANOVA** examples, the personality variables *extra* and *neur* were recoded from the original values (−6 to +6) into three values, using the following **RECODE** command:

```
SPSS/PC: recode neur extra (lo thru 0 = 1) (1 thru 3 = 2)
       : (4 thru hi=3).
```

The command:

```
SPSS/PC: ANOVA var005 BY sex(1,2) neur(1,3).
```

requests a two-way anova with *var005* as the dependent variable. This command produced the output reproduced below.

```
      * * *   A N A L Y S I S   O F   V A R I A N C E   * * *

          VAR005    Women's movement ineffective
     BY   SEX       Subject sex
          NEUR      Neuroticism - coded 3 groups.
```

Source of Variation	Sum of Squares	DF	Mean Square	F	Signif of F
Main Effects	34.096	3	11.365	5.806	.002
SEX	15.488	1	15.488	7.912	.007
NEUR	13.518	2	6.759	3.453	.041
2-way Interactions	1.693	2	.846	.432	.652
SEX NEUR	1.693	2	.846	.432	.652
Explained	35.789	5	7.158	3.657	.008
Residual	84.170	43	1.957		
Total	119.959	48	2.499		

```
     50 Cases were processed.
      1 CASES ( 2.0 PCT) were missing.
```

This table indicates that both of the independent variables *sex* and *neur* have a significant effect on the dependent variable *var005*. The two sexes and the three *neur* groups differ in the degree to which they agree with the statement '**The women's movement has had little serious effect**'. However, there is no significant interaction between *extra* and *neur* with regard to the level of agreement with the statement. From an output of the means for each cell (requested by **OPTION 3**) it was found that the subjects who disagreed most with this statement were non-neurotic females.

Up to five **dependent** variables can be specified on any one subcommand (and five subcommands can be included in a single command – subcommands should be separated by slashes /) e.g.:

```
SPSS/PC: ANOVA var007 to var010 BY neur(1,3)
       : extra(1,3) sex(1,2)
       : /var006 BY neur(1,3) sex(1,2).
```

This would produce four three-way anova tables for the first subcommand and one two-way anova table for the second.

For less than six factors, a full factorial anova is used, with the main effects assessed before the interactions. All interactions are displayed. The anova table includes a count of cases (and a separate count of missing cases), sums of squares, degrees of freedom, the mean square, *F*-

value and the probability of *F* for each of the effects. Variable labels are also included. Missing cases are deleted listwise.

OPTIONS

Altogether, 11 **OPTIONS** are available for the **ANOVA** procedure. The following affect the way in which missing values are handled, the display of labels, the print format and the suppression of interactions.

OPTION 1 specifies that missing values should be included in the analysis.

OPTION 2 requests that labels should not be printed.

OPTION 3 requests that interactions should be suppressed.

OPTIONS 4, **5**, and **6** request the suppression of three-way, four-way and five-way interactions (respectively).

OPTIONS 7 and **8** deal with actions to be taken with regard to **covariates**.

OPTIONS 9 and **10** specify alternative methods (**Regression – OPTION 9**, or **Hierarchical – OPTION 10**) to be used in the analysis of variance. See the *SPSS/PC+ Manual* for details.

OPTION 11 changes the print format to narrow.

STATISTICS

Three additional **ANOVA** statistics are available.

For details on **STATISTICS 1** and **STATISTICS 2** consult the *SPSS/PC+ Manual*.

STATISTIC 3 requests a table of the counts and means for each cell. Thus with two independent variables, one (such as *fac*) specifying three groups, and the other (such as *sex*) specifying two groups, there will be six cells (male arts, female social studies, etc.). The values and counts for the rows and columns of this table are also displayed.

We have now completed our discussion of the procedures **ONEWAY** and **ANOVA**. The use of **ANOVA**, in particular, demands a sophisticated appreciation of statistics, and it should be remembered that anova is a method of experimental analysis rather than simply a statistical test. In a real study, anova should not be used in a *post hoc* fashion, as in the examples given, but should be built into the design of the experiment. As with other methods of analysis, your use of the SPSS/PC+ **ANOVA** procedure should not extend beyond your level of statistical understanding.

13 Statistical Analysis V: Non-parametric Statistics

The **NPAR** procedure allows the user to carry out a variety of non-parametric statistical tests. Unlike parametric statistics (including the **t-test, anova**, the **product-moment correlation**, etc.) non-parametric statistics make few assumptions about the nature of the data. They may be particularly suitable where there are relatively few cases in the population we wish to examine, or where a frequency distribution is badly skewed (rather than symmetrical). Whereas a procedure like **ANOVA** demands that the dependent variable be truly numerical, non-parametric tests can handle data at other levels of measurement. In particular they allow us to analyze nominal and ordinal data.

Nominal data are merely classificatory. For example, players in a sports team will play in different positions. They may even have numbers on their shirts indicating their positions. In a room where members from many teams were assembled we could count the Number 1s, the Number 2s, etc. We might like to know whether significantly more Number 1s than Number 2s were attending the meeting. But it would be quite inappropriate to think of the numbers which indicate the players' positions as real numbers. It would be outrageous, for example, to suggest that 5 Number 1s are in some way equal to one Number 5. The numbers sewn to the players' shirts act only as labels, and cannot be used to add, subtract, multiply or divide. They are merely nominal. Social class and sex are examples of categorical variables. When *class* or *sex* were used, for example, in **ANOVA**, they always appeared as independent variables, never as dependent variables. For some non-parametric tests, however, social class (coded, perhaps, 1 to 5) and sex (coded 1 and 2) could be used as **dependent** variables.

Ordinal data are rank-ordered. The winner of a marathon is first, the runner up is second, and others who finish the race come third, fourth, fifth, etc. The runners may start out with numbers sewn to their shirts, but these are merely labels for identification (nominal). The positions in which they finish, however – first, second, etc. – are not just labels. The numbers mean something. We know that the person who came 31st, for example, must have run a faster race than the person who came 38th. But we **cannot** assume that there was an equal time-gap between the first and the second, on the one hand, and the second and the third, on the other. Neither can we add the numbers (making the second plus the third somehow equal the fifth).

NPAR does not specify a particular test. Fifteen different non-parametric tests are available and the appropriate test must be specifically requested.

13.1 Choosing a non-parametric test

The choice of test depends on a number of factors:

- the nature (level of measurement) of the data (e.g. nominal or ordinal)
- the number of samples or groups (1, 2 or k)
- whether the two (or more) samples are related or independent
- what we want to do with the data (e.g. assess the level of association between two variables, determine differences in the central tendency or span of two or more groups, examine sequence patterns, or examine proportions within a single group)
- the preference of the user (i.e. sometimes a number of alternative tests will be suitable for assessing some aspect of the data)

Table 13.1 presents a concise account of the non-parametric tests available within the **NPAR** procedure. The tests are listed in the order in which they are discussed in the rest of this chapter. The sequence is organized in terms of the number of samples, whether the samples are independent or related, and whether nominal or only ordinal data can be analyzed by the test.

The levels of measurement (nominal, ordinal and, for parametric statistics, interval) can be thought of as occupying different levels on an ascending scale from nominal to interval. A nominal data test (such as chi-square) can be used to analyze data at **any** level. An ordinal test can be used only for data which is **at least** at the ordinal level. The general rule is that a test which is designed to analyze one level of data can be used to analyze data at that level or at any **higher** level, but cannot analyze data which is at a **lower** level.

13.2 One-sample tests

The chi-square test

The chi-square test available within the **CROSSTABS** procedure examines the frequencies of nominal data as they are crosstabulated for the values of two or more variables. **NPAR CHISQUARE**, however, employs the chi-square test in a different manner, to see whether the observed frequencies of data across the values of a single variable conform to an expected pattern.

The output display includes the observed and expected frequencies (with

Table 13.1 Non-parametric tests available within the NPAR procedure

TEST	NPAR NAME	N nominal O ordinal D dichot.	Purpose	No. of samples	Ind. or Related
1 Chisquare	CHISQUARE	N	Tests fit between observed and expected frequencies	1	–
2 Binomial	BINOMIAL	N (D)	Tests for biased proportion	1	–
3 Runs	RUNS	N (D)	Test whether run sequence is random	1	–
4 Kolmogorov-Smirnov	K-S	O	Tests match between frequencies	1/2	I
5 Wald-Wolfowitz	W-W	O	Tests for diffs in ranks of ind. groups	2	I
6 Moses	MOSES	O	Compares ranges of two Indep. groups	2	I
7 Mann-Whitney	M-W	O	Tests for differences between two Indep. groups	2	I
8 McNemar	MCNEMAR	N (D)	Tests before/after change in D variables	2	R
9 Sign	SIGN	O	Tests paired data for + or – bias	2	R
10 Wilcoxon	WILCOXON	O	Test for differences between Related groups	2	R
11 Median	MEDIAN	O	2-way frequency association	2-k	I
12 Kendall	KENDALL	O	Agreement between judges	2-k	R
13 Kruskal-Wallis	K-W	O	1-way anova	k	I
14 Cochran	COCHRAN	N (D)	Tests before/after change in k groups	k	R
15 Friedman	FRIEDMAN	O	2-way anova	k	R

the differences, or residuals), and the chi-square value, with the related degrees of freedom and significance.

All or part of the value-range for the variable can be tested for the fit between the observed and the expected frequencies. By default, equal frequencies are expected for all of the values.

Suppose that we asked 1000 children to choose their favourite flavour of ice-cream from a range of nine flavours. We could code their choice for the variable *prefice* using a 1 to 9 coding system. If we wished to test the assumption that all of the flavours would enjoy equal popularity, then we could rely on the default ('**equal**') and would not need to specify particular expected frequencies. The command used to request a one-sample chi-square test in this case would be:

```
SPSS/PC: NPAR CHISQUARE=prefice.
```

If, however, we had some reason to believe that the strawberry flavoured ice-cream might be twice as popular as any other flavour, we could test the observed preferences against the expected frequencies. In the following example we use a chi-squared test to find out whether the observed frequencies match or fit the expected frequencies of the **five** flavours coded 1 to 5. Thus the example also shows how we can examine the frequency match for only **some** of the values originally coded.

We expect two children to have chosen strawberry flavoured ice-cream (coded 5) for every one child that has chosen **each** of the four other flavours. This could be tested with the command:

```
SPSS/PC: NPAR CHISQUARE=prefice(1,5) / EXPECTED= 4*1, 2.
```

We see from this example that the expected frequencies are interpreted as **proportions** rather than as absolute frequencies. Notice also the convention (in this example **4∗100**) used to specify a uniform expected frequency for a range of values. The expected frequencies must be listed in order of ascending value (in this example, from value 1 to value 5).

NPAR STATISTICS and **OPTIONS** are discussed in Section 13.7.

The binomial test

The binomial test determines whether a variable which takes one of only two possible values (i.e. a dichotomous variable) is randomly (binomially) distributed. By default the test assumes that either of the two possible values is equally likely (i.e. a 0.5 probability is assigned to each value) although this can be changed. The output display includes the count of valid cases, the proportion of cases taking the first value and the two-tailed probability of this proportion. If the probability is low, and reaches significance level, then we can say that the two values of the variable are **not** randomly (binomially) distributed.

A binomial test could be used to determine whether a coin (which can be tossed repeatedly to take a heads or tails value for each toss) is biased – i.e. whether the probability of the observed heads and tails frequencies differs significantly from 0.5. Using the stereotype variables from the gender attitudes survey (*st01* to *st10*) we can determine whether the proportion of respondents who felt that more men or more women would agree with each of the 10 statements was significantly different from a 50:50 split.

A binomial test to determine whether *st05* was biased could take the form:

```
SPSS/PC: NPAR BINOMIAL=st05 (1,2).
```

Because each of the stereotype variables takes only two valid values, no values need be specified. A list of variables can be included. The following command illustrates both of these points:

```
SPSS/PC: NPAR BINOMIAL=st01 to st10.
```

The output from this command included the following:

```
- - - - - Binomial Test

    ST01        Depressed often (M-W)

    Cases
       48    =   2            Test Prop. =  .5000
        1    =   1            Obs. Prop. =  .9796
       --
       49    Total           Z Approximation
                             2-tailed P =   0.0
```

From this we can conclude that the values for this variable are not randomly distributed between the two possible values. We can see that the respondents believed that more women than men would agree with the statement: '**I often feel a little depressed**'.

The results of the above command also revealed that there was a significant stereotyping effect on several of the other variables.

Further results indicated that women were also believed to be more likely than men to report feeling anxious and having sleep difficulties. Men, on the other hand, were seen as more likely to report thinking about sex a lot, enjoying taking risks and solving abstract problems, and trying to dominate people.

The binomial test can also be used when the prior probability of a particular value is not 0.5. The following example tests the hypothesis that a die is biased such that the value 1 is presented with a frequency significantly greater than 1/6 (0.167). The results of successive throws are coded 1 to 6 for the variable *throw* and the frequency of 1s is tested against the perfectly unbiased expected frequency:

```
SPSS/PC: NPAR binomial(.167)= throw(1).
```

The 1 in brackets after the variable name specifies that the value 1 is the first value of the dichotomy and that **all other values** (in this case 2 to 6) are treated together as the second value (which would have the expected frequency of 0.833, or 83.3 per cent).

If the default probability (0.5) is changed then the output gives the **one-tailed probability**.

The runs test

The runs test is used to determine whether, for a dichotomized variable (such as heads and tails, or male and female) the **sequence** of values across cases is random. Thus we could test whether, in a two-party state there is a tendency for one party to remain in office through several successive elections, until the other party is elected (after which it, too, stays for several terms). We could test this by analyzing historical data, where each case is a term in power (the variable *govmnt*) and the two parties are coded 1 and 2. Clearly, the order in which the cases were placed would be of the utmost importance (in this example it would have to be the chronological order of the successive governments). The appropriate **NPAR RUNS** analysis could be requested using the following SPSS/PC+ command:

```
SPSS/PC: NPAR RUNS(2)=GOVMNT(2).
```

The (2) specifies the cutting point. Values **below** this are treated as being on one side of the dichotomy, and values equal to or greater than this point are treated as being on the other side.

The display includes the cutting point, the count for each side of the dichotomy, the Z statistic with its one-tailed probability, and the number of runs – a run is an unbroken sequence of one value. Thus, in the examples below, list (a) contains 8 runs and list (b) contains 21 runs:

(a) 00000110111111101111111111101111111111
(b) 001101000011101010011101110110101110

Variables can be dichotomized either by specifying a numerical cutting point, as in the command example above, or by ordering that the cutting point be at the mean, the mode or the median, e.g.

```
SPSS/PC: NPAR RUNS(median) = test1.
```

In the following example, nine variables are subjected to the **RUNS** test in three subcommands:

```
SPSS/PC: NPAR RUNS(mode)=item1 item9/ RUNS(mean)= item2 to
       : item6/ RUNS(18.99)= score4 score7.
```

13.3 One- or two-sample tests

The Kolmogorov–Smirnov tests

The Kolmogorov–Smirnov (K–S) test is used either to compare the frequency distribution of one sample against a theoretical distribution (the **one-sample test**) or to compare the frequency distributions of two independent samples (the **two-sample test**).

One-sample test

The one-sample K-S test is a test of the goodness of fit between a distribution of observed frequencies and a theoretical frequency distribution. The theoretical distributions available as templates for matching are the *normal* distribution, the *Poisson* distribution and a *uniform* (or rectangular) distribution – i.e. equal frequencies for all values. The value computed is Z, a measure of how much the observed distribution differs from the theoretical.

By default, the display includes the count of valid cases, the range, the Z value (with its two-tailed probability) and the most extreme differences between the observed and theoretical frequencies.

The theoretical distribution with which the observed frequencies are to be compared must be specified in brackets after the K-S characters (e.g. **(normal)**).

- for the **normal** distribution a **mean** value of the theoretical distribution and its **standard deviation** can be specified
- for the **Poisson** distribution a **mean** can be specified
- for the **uniform** distribution **minimum** and **maximum** values can be specified

Such additional parameters are optional (by default these values are taken from the observed frequency distribution).

The following example tests whether the observed frequencies of grades (from A to L) given to students (the variable *grade*, coded 01 to 12) conform to a **uniform** distribution (i.e. approximately equal numbers of students receive each of the 12 grades):

```
SPSS/PC: NPAR K-S(UNIFORM)=GRADE.
```

The next example tests whether the frequency distribution of IQ scores obtained from a population of children in a large national survey conforms to the theoretical distribution – a **normal** distribution with a mean of 100 and a standard deviation of 15:

```
SPSS/PC: NPAR K-S(NORMAL, 100,15)=IQ.
```

A non-significant probability value in the output display indicates that that distribution does **not** depart significantly from a normal distribution with this mean and standard deviation.

You could use the command:

```
SPSS/PC: NPAR K-S(NORMAL)= extra neur sexism.
```

to test whether the frequencies of the values for any of these (gender attitudes survey) computed variables (**not** recoded into categories) depart significantly from a normal distribution (with mean and standard deviation unspecified, and therefore taken from the observed distributions).

Two-sample test

The two-sample Kolmogorov–Smirnov test compares the distribution of frequencies for two observed groups defined by an **independent** (grouping) variable. Thus it could be used to compare the frequency distribution of answers to a seven-point scale question for, say, males and females. The data used should not be categorical (i.e. it should be at least at the ordinal level).

Such a command would take the following form:

```
SPSS/PC: NPAR K-S= var001 BY sex(1,2).
```

To compare the *sex* distribution for a number of variables a list could be used:

```
SPSS/PC: NPAR K-S= var001 to var003 BY sex(1,2).
```

This would give an analysis of the similarity of the distribution for the two sexes for each of the three variables specified. A non-significant probability would indicate that the distributions of the answers for the two sex groups do not differ significantly.

The display includes the count of valid cases, the **K-S** *Z* value (with its two-tailed probability) and the most extreme differences between the two groups.

13.4 Two-sample tests

The Wald–Wolfowitz test

The Wald–Wolfowitz (W–W) test rank-orders the scores from two **independent** samples (for a dependent variable) from low values to high values as if they came from a single population. A runs test analysis is then performed on the ranks in order to determine whether the dependent variable is **distributed** evenly thoughout the ranked sequence. If the two samples differ substantially in their central tendency, their variability or their skewness, then a significant *Z* value will be obtained.

If boys (B) and girls (G) of a certain age were rank-ordered for height, for example, the sequence:

BGGBGBGBBBGGGBBGGGBBBBGBGBGBBBGGBG

would suggest that height was not strongly associated with sex (i.e. heights were not different between the sexes). The sequence contains 20 runs. However, the sequence:

BBBGBBBBBBGGBBBBGBGGGGGBGGGGGGBGGGG

might suggest that boys were shorter than girls (there are 12 runs).

Each display includes the count for each group, the number of runs (or the minimum and maximum number of runs if there are ties between the groups). A Z value is given for the number(s) of runs, together with the one-tailed probability.

Suppose that we have the idea that males make more facial movements or hand gestures when watching violent films than females. We could count the number of such movements (the variable *moves*) as groups of men and women watch a 15-minute presentation of violent film extracts. An appropriate SPSS/PC+ command for analyzing the data with a Wald–Wolfowitz test would be:

```
SPSS/PC: NPAR W-W= moves BY sex(1,2).
```

The Moses test

The Moses test of extreme reactions (for two **independent** samples) examines whether the **range** of an ordinal variable is similar for the two groups. One of the groups acts as the control group, and the other is the comparison group.

The Moses test is useful when we suspect that one group of subjects has extreme reactions (i.e. both low and high values) compared to another group. Thus the two groups might have similar means or similar mean ranks, but their span might be significantly different. Data from both groups is rank-ordered in a single ascending sequence and the span of the control group is found. From the size of the total sequence the probability of such a span occurring by chance can be calculated. To override distortions which can arise from presence of the odd extreme value in the control group, a degree of trimming can be implemented. By default the extreme 5 per cent of cases from each end (known as outliers) are trimmed in this way.

The display includes the counts of valid cases in the two groups, the span of the control group (with and without the outliers) and the associated one-tailed probabilities.

Suppose we believe that when passing a poster proclaiming the health dangers of smoking, smokers tend either to pass quickly on or to examine the contents in detail (compared with non-smokers). We could measure the time taken to pass by the poster (*passtime*) for both smokers (coded 1 on the variable *smoker*) and non-smokers (coded 0) and then subject our hypothesis to the Moses test with the following SPSS/PC+ command:

```
SPSS/PC: NPAR MOSES=passtime BY smoker(0,1).
```

The order of the values specified for the variable *smoker* determines which group is treated as the control group and which is treated as the experimental group. The first value specifies the control group (in this cases non-smokers, value 0). We might decide to trim 10 per cent of control group outliers from either end of the rank sequence. With, say, 200 non-smokers as controls we would do this by specifying that 20 non-smokers should be trimmed from both ends, using the following command:

```
SPSS/PC: NPAR MOSES(20)=passtime BY smoker(0,1).
```

The Mann–Whitney test

The Mann–Whitney (M–W) test is a two-sample test which compares the ranks of two **independent** groups defined by a grouping variable. (It is therefore in many ways similar to the independent *t*-test procedure for parametric data.) The values of the dependent variable are rank-ordered for both groups as if they came from a single population. The test statistic *U* is calculated in a formula which includes *n*1 and *n*2 (the numbers in the two groups) and the larger of the two totals of ranks for the groups. The **lower** the value of *U* the more significant are the ranks of the two conditions (groups).

The output display includes, for both groups, the valid count and the mean rank of the variable. If there are fewer than 30 cases then the *U* statistic is given with its exact two-tailed probability. If there are more cases then *U* is converted to a *Z* statistic. The Wilcoxon *W* value is also given (see later), together with its probability.

The **M-W** subcommand could be used, for example, to determine whether there was a significant sex difference in the number of calories which subjects said they consumed in an average day – the variable *cals*.

A suitable command would be:

```
SPSS/PC: NPAR M-W=cals BY sex(1,2).
```

This command would cause the system to rank-order cases in terms of their value for *cals* and then to sum the ranks for each of the conditions (the male condition and the female condition). The higher sum of ranks would then be used to calculate the value of *U*.

The McNemar test

The McNemar test is used to compare two dichotomous **related** variables which have been coded with the same two values. If the variables to be tested are not already in this form then the data should be **RECODED** and the recoded variables used for **NPAR MCNEMAR**. The test is particularly useful for before-and-after designs where one variable represents a dichotomy before an event or manipulation and the other variable represents the same dichotomy after the event or manipulation. The data can be nominal.

Suppose that we wish to know whether voting intentions (in a two-party state) have changed significantly since a particular event (for example, a scandal involving a prominent politician). If we have a record of subjects' voting intentions **before** the scandal broke (*vote1*), **immediately after** the scandal broke (*vote2*), and some **three weeks after** the scandal broke (*vote3*) then we can analyze the changes which have taken place in voting intentions.

The display presents a 2×2 table with variable names (or labels) and values, the total number of valid cases and the two-tailed binomial probability.

The simplest form of the command would be:

```
SPSS/PC: NPAR MCNEMAR = vote1 vote2 vote3.
```

This would produce three separate McNemar tests, because each **pair** of variables included in the varliable list would be tested together (in this case *vote1* with *vote2*, *vote1* with *vote2* and *vote2* with *vote3*).

The keyword **WITH** can be used. Thus:

```
SPSS/PC: NPAR MCNEMAR = vote1 WITH vote2 vote3.
```

would produce just **two** tests (*vote1* with *vote2*, and *vote1* with *vote3*).

A variable list can be included on either side, or both sides, of the keyword **WITH** (see also the section below on **OPTIONS**).

A McNemar test can be used to compare the 'ideal date' and 'ideal marriage partner' data (*dat01* to *dat08* and *mar01* to *mar08*) from the gender attitude survey. This data can be recoded so that the 'very important' rating retains the value 1 and all other responses take the value 2:

```
SPSS/PC: recode dat01 to dat08 mar01 to mar08 (2,3,4, =2).
```

The McNemar test can now be used to see whether the frequency of the 'very important' rating is different for ideal dating and marriage partners for each of the eight attributes included:

```
SPSS/PC: npar mcnemar dat01 to dat08 with mar01 to mar08
       : /option = 3.
```

OPTION 3 is specified here to request a special **WITH** pairing (as for the **T-TEST**). This means that *dat01* will be compared only with *mar01*, that *dat02* will be compared only with *mar02*, etc.

The output from this command included the following:

```
- - - - - McNemar Test

        DATØ8        Dependability (date)
  with MARØ8         Dependability (spouse)

                      MARØ8

                     2             1         Cases       43
               I_____I_____I
            1  I           1 I         17  I
  DATØ8        I_____I_____I    (Binomial)
            2  I          15 I         10  I    2-tailed P   .0117
               I_____I_____I
```

(1 is 'very important'; other ratings are represented by 2)

Here we see that there is a significant difference in the frequency of those who stress dependability as a 'very important' characteristic of a date and of a spouse. Ten respondents felt that dependability was a 'very important' characteristic of an ideal spouse but not of an ideal date. For only one subject was the reverse true.

The sign test

The sign test examines pairs of data (i.e. it compares the values of two specified variables for each case) to test whether the number of positive differences between the variables (variable A greater than variable B) is the same as the number of negative differences (variable B greater than variable A). Clearly such comparisons mean that data must be able to be ordered, so data subjected to the sign test must be at least at the ordinal level of measurement.

The display provides counts of the number of positive and negative differences, the number of ties (i.e. no difference between the two values), and the two-tailed binomial probability of the sign distribution obtained.

Suppose we are interested in whether subjects who have previously taken a general knowledge test (without being told whether the answers they gave were right or wrong) are as likely to get a lower score when tested again as they are to get a higher score. If the two scores obtained give the values for the variables *test* and *retest*, the following **SIGN** procedure could be used to determine whether subjects were equally likely to get a higher or lower score on retest:

```
SPSS/PC: NPAR SIGN=test retest.
```

The keyword **WITH** can also be used, with variable lists on either or both sides of the keyword (this is requested with **OPTION 3** – for further details of the **NPAR OPTIONS** see the end of this chapter).

The Wilcoxon test

The Wilcoxon test (or Wilcoxon signed-ranks test) is used to assess whether two **related** samples (with data at the ordinal level or above) are significantly different (it is in many ways equivalent to the matched-pairs *t*-test procedure for parametric data). The differences between the pairs of variables are ranked for magnitude, irrespective of whether the differences are positive or negative. The sums of the ranks are then determined separately for those cases in which the difference is positive and those cases in which the difference is negative, and the smaller of these sums is the value *W*. From the number of pairs and the value of *W*, another statistic, *Z*, can be calculated and its probability determined.

The **NPAR WILCOXON** output display includes the total valid count and the mean rank for both variables. The number of positive and negative rank differences and the number of tied ranks is also given. The *Z* value is included together with its two-tailed probability.

Suppose that we wish to test the effectiveness of a drug which is claimed to improve memory performance. We could test subjects' normal memory (*normem*, without the drug) and then the same subjects' memory with the drug (*drugmem*). We wish to answer the question, 'Is there a significant difference between the paired values (one pair per case) of the two variables **normem** and **drugmem**?'

The following **NPAR** command could be used to determine whether there was such a difference:

```
SPSS/PC: NPAR WILCOXON = normem drugmem.
```

The keyword **WITH** can also be used, with variable lists on either or both sides of the keyword. (For further details on specifying the comparisons to be made see Section 13.7 on **OPTIONS**.)

13.5 Two- or k-sample tests

The median test

The median test is a test for **independent** groups for which data is available at the ordinal level of measurement. It can be used to compare either two groups (the **two-sample median test**) or more than two groups (the **extended median test**), and determines whether there are differences between groups in terms of their central tendencies. For both versions of the test the data is cast into a $2 \times k$ crosstabulation, where each of the two rows represents a split of values at the overall median (or some other specified cutting point) and each of the k columns represents a particular group.

The two-sample median test

A 2×2 table is formed and appears in the display. The overall median (for the two groups combined) is calculated and used to cut the values for each group into two sections. The first row contains the frequency for each group of values which are greater than the median. The second contains the frequencies of the values which are less than or equal to the median. The first column contains the data for the first specified group and the second column contains the data for the second specified group. The display also includes the number of valid cases, the median, and the chi-square statistic with its significance.

Suppose that we have asked a group of subjects to list their phobias (fears of spiders, snakes, heights, confined spaces, etc.). The data for the variable *fears* (the number of phobias each person has listed) may be highly skewed, with many subjects reporting few fears and some reporting many. To test whether there is a significant difference in the number of fears reported by men and by women we could use the median test. The variables used would be *fears* and *sex*.

An appropriate SPSS/PC+ **NPAR** command for requesting a two-sample median test would be:

```
SPSS/PC: NPAR MEDIAN= fears BY sex(1,2).
```

Here the cutting point would be the overall median of the two groups combined. Alternatively, we could specify a cutting point:

```
SPSS/PC: NPAR MEDIAN (3)= fears BY sex(1,2).
```

Cases with more than three fears would be assigned to the top row of the crosstabulation. Cases with three or fewer fears would be assigned to the bottom row.

The extended median test

The median test can also be used to determine whether there are overall differences for more than two groups (k groups). The data is cast into a $2 \times k$ crosstabulation, with the rows again representing a split of values at the overall median (or some other specified cutting point) and each of the k columns representing a particular group. From this table a chi-square statistic is calculated. The display is essentially the same as that of the two-sample test, but the number of degrees of freedom of the table is also reported.

Suppose that subjects are assigned to three anxiety level groups on the basis of a questionnaire (high anxiety, medium anxiety and low anxiety, coded 1, 2 and 3 respectively on the variable *anxiety*). If we now wish to determine whether there are differences between these groups in terms of the number of specific fears or phobias they report (counted on the variable *fears*), we could use the following SPSS/PC+ command:

```
SPSS/PC: NPAR median=fears BY anxiety(1,3).
```

This would give a chi-square statistic for a 2×3 table in which each of the three groups was included, with a split at the overall median.

On the other hand, reversing the parameters specified for the variable *anxiety*:

```
SPSS/PC: NPAR median=fears BY anxiety(3,1).
```

would yield a two-sample median test with only groups 3 and 1 included.

The Kendall test

The Kendall test which is available within the SPSS/PC+ **NPAR** procedure calculates *W*, the coefficient of concordance. This is a measure of the degree to which a number of judges (two or more) agree in their rating (or ranking) of a number of items. *W* has a range from 0 to 1, where 0 means that there is no agreement between the judges and 1 means that there is total agreement. In more general terms, the Kendall test assesses whether two or more (*k*) **related** samples could be from the same population.

The display includes the mean rank (across all judges) for each variable, the total number of judges (here cases), *W*, the chi-square equivalent, degrees of freedom (= number of variables − 1) and the significance value.

Suppose that five judges have been called upon to rate 12 photographs (variables *photo1* to *photo12*) on a 10-point scale of, say, attractiveness. To assess the overall agreement between the judges we could use the following SPSS/PC+ **NPAR** command:

```
SPSS/PC: NPAR KENDALL = photo1 to photo12.
```

13.6 k-sample tests

The Kruskal–Wallis test

The Kruskal–Wallis (K–W) test is a non-parametric one-way analysis of variance and is similar to the Mann–Whitney test (except that three or more conditions may be involved). The test is used to establish whether there is a significant overall difference in a dependent variable for a number of groups differentiated on a particular independent variable. It is therefore an **independent** samples test. The data must be at least ordinal. The data are ranked (all as one population) and the rank-sum for each group is calculated. The measure obtained is H.

The display includes the mean rank and the count for each group, the number of valid cases, the chi-square equivalent of H, and its significance level. Where there are tied ranks across groups the chi-square value corrected for the ties is also displayed, again with the appropriate significance level.

Suppose that we wish to discover whether there are differences in the attitude towards capital punishment (*attcp*) between groups of subjects who have been given one of three relevant reports to read. One report is heavily biased in favour of capital punishment, one is dispassionate and essentially neutral, and the other is highly critical. The variable *report* is coded according to which report has been read and the three reports have been assigned the values 1, 2 and 3. *Report* is the independent variable and *attcp* the dependent variable.

The **NPAR** command used to request a Kruskal–Wallis one-way analysis of variance for this study could be:

```
SPSS/PC: NPAR K-W = attcp BY report(1,3).
```

This command creates three *report* groups (the numbers in brackets indicate the minimum and maximum values to be used in forming the groups). The dependent variable *attcp* is ranked across all groups and the sums of the ranks for each group are then compared.

The Cochran test

The Cochran test can be regarded as an extended form of the McNemar test. It assesses whether, for several **related** samples, the proportions assigned to a particular (dichotomous) value (nominal or ordinal) are significantly different. Thus if, in a series of test questions, **similar** proportions of subjects supply the correct answer to each question, the Cochran test statistic, Q would not be significant. If there were substantial differences between these proportions, however, then Q **would** be significant.

The test involves the construction of a $2 \times k$ crosstabulation (where each of the two rows contains the count for one or other side if the dichotomy, and k is the number of variables).

The display gives the count for each of the two values for each variable, the total valid count, and the Cochran Q statistic with the appropriate degrees of freedom and significance value.

Suppose that we have respondents' answers to eight general knowledge questions (*Q1* to *Q8*) and we wish to establish whether the difficulty level of the questions is approximately equal. The answers could be scored as 'right' or 'wrong' and the Cochran test used to determine whether, overall, there was a significant range of difficulty within the set of items.

A suitable SPSS/PC+ command for requesting such a Cochran analysis would be:

```
SPSS/PC: NPAR COCHRAN = Q1 to Q8.
```

The Friedman test

The Friedman test can be regarded as an extension of the Wilcoxon test. It is a two-way analysis of variance by ranks which tests whether two or more **related** samples (variables) have been drawn from the same population. If the chi-square value produced is significant then it can be concluded that the variables have not come from the same population.

Data from a number (n) of cases are available for each of a number (k) of conditions. The data must at least be at the ordinal level of measurement. The data are cast into a $n \times k$ table. In the Friedman analysis, the data from each case are ranked **horizontally** from 1 to k and a two-way analysis of variance (cases by conditions) is performed. If some conditions (variables) have a substantially lower (or higher) sum of ranks than others then this indicates that the conditions are significantly different.

The **NPAR FRIEDMAN** output display includes the mean rank for each of the variables (conditions), the degrees of freedom for the table, and the chi-square value with its significance.

Suppose that five children have each been asked to rank-order eight flavours of ice-cream (the variables *ice1* to *ice8*) in terms of their own preference. The data matrix in Table 13.2 has emerged (1 means ranked first and 8 means ranked last).

Table 13.2

Child	Ice-cream flavours (k)
	1 2 3 4 5 6 7 8
001	8 6 1 3 5 7 2 4
002	7 5 3 2 6 8 1 4
003	6 7 5 1 8 4 2 3
004	8 4 2 3 7 6 1 5
005	8 7 2 3 6 5 1 4

The data given in the body of the table are the actual ranks as chosen by the children. If the children had rated the ice creams instead (e.g. given them points from 1 to 20) then SPSS/PC+ would have ranked the data before performing the Friedman test.

To obtain a Friedman analysis, the appropriate command would be:

```
SPSS/PC: NPAR FRIEDMAN= ice1 to ice8.
```
With the above data the following display was obtained:

```
- - - - - Friedman Two-way ANOVA

    Mean Rank   Variable        Cases       Chi-square
        7.40    ICE1              5           28.2000
        5.80    ICE2
      ...                        D.F.       Significance
        6.00    ICE6              7             .0002
        1.40    ICE7
        4.00    ICE8
```

13.7 NPAR: STATISTICS and OPTIONS

Additional statistics may be requested. In many cases, however, only some of these additional statistics will be valid or meaningful.

STATISTIC 1 requests that the mean, maximum, minimum, standard deviation and count of each variable should be displayed.

STATISTIC 2 requests the display of the count and of quartiles (i.e. the values which provide the cutting points for 25 per cent, 50 per cent and 75 per cent of the frequency distribution for each variable).

By default, variables with missing values are deleted from analyses of single samples. For two or more samples they are deleted pairwise.

OPTION 1 requests that missing values should be **included** in the analysis.

OPTION 2 requests the listwise deletion of cases with missing values. A case with a missing value for any variable listed on a subcommand is excluded from all analyses requested by that subcommand.

OPTION 3 concerns the pairing of variables in subcommands requesting analysis with **two related** samples (it can therefore be used only with three **NPAR** tests – McNemar, sign and Wilcoxon. This option may be specified with or without the keyword **WITH**. There are thus four possibilities, as shown in Table 13.3.

OPTION 4 applies only to situations in which the size of the data file to be analyzed may be too great for the available computer memory. If option 4 has been specified then random sampling of the data is executed if there is insufficient memory.

Table 13.3

WITH included?	Option 3 specified?	
No	No	Every variable paired with every other
Yes	No	Each variable to left of WITH paired with each variable to the right of WITH
No	Yes	Sequential pairs tested (i.e. 1 with 2, 3 with 4, 5 with 6, etc.)
Yes	Yes	1st variable before WITH paired only with 1st after, 2nd with 2nd, etc.

14   Reporting Findings

When the results from a study have been analyzed, the next task is to produce a report of the findings. Several features built into the SPSS/PC+ system provide help at this stage. For example, output from the various SPSS/PC+ procedures, saved in the form of named listing files at the end of analysis sessions, can be easily edited and incorporated into word-processing files. The text of a report can therefore be written around SPSS/PC+ output – there may be no need for the results to be retyped.

The cost of SPSS/PC+ software, and the fact that the program is run on a hard-disk machine, might mean that only one or two running systems will be available within a particular department or computer centre. If there is a high pressure on the use of the SPSS/PC+ resource, a bottleneck may develop. One way of dealing with this problem is to prepare work for use by SPSS/PC+, and later to produce reports based on SPSS/PC+ output, using an ancillary microcomputer. The following scenario depicts one way in which highly efficient use can be made of a limited SPSS/PC+ resource:

- An ancillary (or off-site) microcomputer running a suitable editor program (a word-processor, for example) can be used to create a data file, a data definition file, and a set of command files.
- A diskette containing these files can then be copied into the appropriate directory of the SPSS/PC+ microcomputer.
- SPSS/PC+ can be used to **INCLUDE** the definition file, and the editor program **REVIEW** used to correct errors.
- The various command files can now be **INCLUDED** to perform a series of operations, transformations and precedures. If errors emerge, **REVIEW** can be used to re-edit the relevant files.
- The listing file which is automatically saved by SPSS/PC+ at the end of the session can be copied at the end of the session onto a diskette and, if necessary, transferred to another microcomputer.
- Extracts from the listing file can now be edited on the word-processor and saved (perhaps as a number of small files) for later incorporation into the text of the report.

Using such a strategy, well over 90 per cent of the work-time would typically be spent on a microcomputer other than that used to run SPSS/PC+ and thus many more users could make use of the SPSS/PC+ facility.

The microcomputer used for file preparation and word-processing does not need to be directly compatible with the IBM machines for which SPSS/PC+ is designed. A number of programs are now available for transforming text files (and data files) to and from the IBM format. Any one of a very wide range of microcomputers could therefore be used as the principal machine for a project to be analysed by SPSS/PC+.

14.1 The REPORT procedure

One SPSS/PC+ procedure, **REPORT**, provides a particularly useful aid to report-writing, allowing the user to produce tables of summary statistics after the cases in the population have been broken down by one or more grouping variables.

A **REPORT** command can be used to produce a listing of the values of individual cases, or it can be used to produce a table of summary statistics for different groups of cases. The following discussion will focus on this latter use.

The **REPORT** procedure displays summaries of the data for one or more variables after cases have been broken down by break variables. The output table is arranged so that the **columns** (not equivalent to the single-digit columns used in data entry) contain information about particular **variables**, and the **rows** correspond to **groups** of cases. **REPORT** provides a flexible means of tailoring the presentation of the table (in terms of labelling, titling, column width, and spaces between lines). Although **REPORT** is worth exploring, it should be realized that there are several other ways of producing essentially the same result. Thus the output from other procedures (especially **MEANS**) will produce output which may be similar to that obtained from **REPORT**. A listing file obtained from the output of the **MEANS** procedure can be edited with a word-processor to produce useful summary tables. Also, one of the optional SPSS/PC+ modules, **TABLES**, has been specifically designed to produce high-quality tables. Many users may therefore prefer not to use **REPORT** but will rely instead on one of the other methods of producing report tables.

Because **REPORT** is flexible, it is also somewhat complex. It requires several subcommands, and many optional subcommands are also available. It also assumes that the cases within the dataset are arranged in a suitable sequence. Thus a **SORT** command is usually issued before the **REPORT** procedure is executed (see below).

The following example illustrates the form of a complete **REPORT** command (examine it, but do not try to understand the various subcommands at this stage):

```
SPSS/PC: sort by sex.
SPSS/PC: report variables = extra "extrav" neur "neurot"
       : /break = sex (label)
       : /summary = mean "Average"
       : /summary = stddev "S.D.   ="  (neur)
       : /title "Summary Statistics of E, N and Sexism".
```

To produce a summary table, a **REPORT** command must include the subcommands **Variables**, **Break**, and **Summary**.

The **Variables** subcommand specifies the variables which are to be reported in the table, e.g.

```
/variables = var001 to var003
```

The **Break** subcommand specifies how the **cases** are to be broken down into groups, e.g.

```
/break = sex
/break = sex fac
```

The first example would break the cases into two groups, each with one of the values for the variable *sex*. The second example would break the cases first into *sex* groups and then, within each sex, into *fac* sub-groups. This would produce a two-break report. The file would have to be pre-sorted on **both** of the break variables. If the two break variables had been specified in the reverse order then the cases would have been broken first into *fac* groups and then, within these groups, into two *sex* sub-groups.

The **Summary** subcommand lists a statistic which is to be calculated for each group and displayed in the table, e.g.

```
/summary = mean
```

By default the specified summary statistic is displayed for all variables. If a statistic is to be provided for only some of the variables then these can be specified. Thus:

```
/summary = mean (neur)
```

This would produce a display in which the mean value was given only for the named variable.

A **Format** subcommand can also be used to change defaults for spacing between lines and the size of the output page (see the *SPSS/PC+ Manual*, pp. C148–150, for details).

Title and **Footnote** subcommands can be used to specify various forms of labelling for the table.

No **Statistics** or **Options** subcommands are available for use with **REPORT** (although a variety of summary statistics can be requested on a **Summary** subcommand).

Subcommand order

The order in which subcommands are entered is important:

- If a **Format** subcommand is included, it should be specified **first**
- **Variables** should **follow** the **Format** subcommand (if there is no **Format** subcommand it should be placed first)
- The **Break** subcommand must be placed **after Variables**
- **Summary** must follow **immediately after Break**
- **Title** (and **Footnote**) can be placed anywhere except between a **Break** subcommand and a **Summary** subcommand which immediately follows it.

Labelling the table

Within various **REPORT** subcommands there are opportunities for requesting that existing labels be added into the table, and new labels can also be provided at this stage.

Thus within the **Variables** subcommand a heading can be provided for any column that will contain the summary statistics for a variable. The label provided should be short enough to fit within the width of the column assigned to that variable within the table (the default width is eight columns). If necessary, a label can continue over a number of rows, with material to be entered into each row specified inside a separate pair of quotation marks. Thus:

```
variables = var001 "Women" "equal to" "   men"
```

would produce the heading

```
Women
equal to
    men
```

If no column title is specified then the existing variable label is used (or, if no label has been provided, the variable name).

For the **Break** variables, the default labelling is simply the value (thus if 1 had been used for males, the 'male' row would be labelled 1). However, if **(label)** is specified then the existing value label is used.

Thus a subcommand of the form **/break sex** would produce a display which included the values (e.g. 1 and 2) whereas a subcommand of the form **/break sex (label)** would produce a display which included the value labels (e.g. 'male' and 'female').

The **Summary** subcommand is used to specify summary statistics which should be included for each variable within the table. Thus:

```
(a) : /summary = mean
    : /summary = sum
```

or

   (b) : `/summary = mean sum`

The first of these examples would cause each of the two summary statistics to be printed on a separate line of the table. The second example would cause both statistics to be presented on the same line.

A selection of the statistics which can be specified by a **Summary** subcommand is presented in Table 14.1.

Table 14.1

Report summary statistics	
validn (valid number of cases)	sum
min	max
mean	stdev
variance	kurtosis
skewness	median
mode	
PCGT(n)	(percent of cases with a value greater than n)
PCLT(n)	(percent of cases with a value less than n)
PCIN(min,max)	(percent of cases with a value between the two values specified)

In addition to the simple summary statistics presented in the table, two aggregate functions (e.g. **abfreq(min,max)** – the **frequency** of cases for **each** of the values within the range specified) and eight composite functions (e.g. **add (a b c)** – the addition of two or more specified variables or summary statistics) can be requested. For fuller information, consult the *SPSS/PC+ Manual*, pp. C156–C159.

By default, the title given to a summary line is the keyword of the function specified for that line (e.g. '**mode**'), but an alternative summary title can be provided in the following way:

```
/summary mean "Average = "
/summary validn  "Count "
```

Organizing the data before issuing a REPORT command

REPORT assumes that the cases within the dataset have been appropriately organized **before** the procedure is implemented. Failure to do this will result in a very odd output table. If a **REPORT** were requested on the variable *extra* (from the gender attitudes survey), for example, with *sex* specified as the break variable, then summary statistics relating to *extra* would be provided for 33 different 'sex groups'. This is because the system would read the cases sequentially within the file and create a new group whenever the *sex* value changed (from 1 to 2, or from 2 to 1).

To prevent this, a prior **SORT** command must be issued to allow **REPORT** to employ the **Break** subcommand appropriately. Thus with the file re-ordered by *sex*, only one change of values for the *sex* variable would be encountered as the system read through the cases, and summary statistics would therefore be provided for just two *sex* groups. Before requesting a two-break report, the **SORT** command should specify a re-ordering of cases **BY both** variables (in the appropriate order). For example:

```
SPSS/PC: SORT by sex fac.
```

This would rearrange the cases within the active file so that all the male cases precede the female cases. Within each sex the cases would be ordered by *fac*.

By default, the output table includes any summary statistics requested, but does not list the values for individual cases. Formatting (margins and page size) follows any instructions issued in a previous **SET** command. If no relevant **SET** commands have been issued then single blank lines are inserted between the title, column headings, break headings and summary lines, etc. These spacings can be changed by an appropriate **Format** subcommand. The default column width for each variable is eight single columns. The break groups are labelled, by default, with the coding values and the columns are labelled with the previously specified variable labels, if such have been provided, or with the variable names. Summary statistics are labelled with the keyword used to specify them on **Summary** subcommands. Cases with missing values are included in a listing of individual case data but are excluded from all calculations.

14.2 Examples

Two simple examples will be given of **REPORT** commands used to produce tables of results from the gender attitudes survey.

Example 1

In the first example, summary statistics (mean and standard deviation) for three of the computed variables (*extra*, *neur* and *sexism*) are to be displayed for the male and female groups of subjects. Thus the variable *sex* is the break variable, and the **REPORT** command is preceded with a simple **SORT** command:

```
SPSS/PC: sort by sex.
SPSS/PC: report variables = extra "extrav" neur "neurot"
       : sexism
       : /break = sex (label)
       : /summary = mean "Average"
       : /summary = stddev "S.D.   ="
       : /title "Summary Statistics of E, N and Sexism".
```

The **Variables** subcommand specifies the variables to be included in the output table. The **Break** subcommand identifies the break variable as *sex* and specifies that the groups formed by this break should be labelled with the value labels previously specified for *sex*. Two summary statistics are requested: the mean (for which the label 'Average' is provided), and the standard deviation (for which the label 'S.D. =' is provided), and these will be printed on two separate lines. The **Title** subcommand provides a title to be printed at the head of the table.

The last command produced the output table shown below:

```
          Summary Statistics of E, N and Sexism

        Subject              extrav       neurot       SEXISM
           sex


Male

Average                      1.8182       2.0000      27.1905
S.D.   =                     2.9703       3.2219       5.5733

Female

Average                      1.8929       2.2500      21.9259
S.D.   =                     2.7126       2.9892       5.0301
```

Example 2

This example displays summary statistics for some of the self-report personality variables (*prs01* to *prs03*) following a two-break analysis by *sex* and *fac*.

First, the appropriate **SORT** command is issued:

```
SPSS/PC: SORT by sex fac.
```

This reorders the file so that all males (coded 1) are placed before all females (coded 2). Within each sex group the cases are then ordered by *fac*. Thus the order will be arts (coded 1), science (coded 2) and social studies (coded 3).

A report table is now requested with the following **REPORT** command:

```
SPSS/PC: report variables= prs01 "Often " "depressd"
       : prs02  "  Sex  " "thoughts" "often"
       : prs03  " Often " " anxious"
       : /break= sex "Subject"  "  sex" (label)
       : /break= fac "Home   " "faculty" (label)
       : /summary = pcgt (1.5) "% FALSE".
```

Three variables are to be included in the table. For each of these, a suitable column heading has been provided. Two break subcommands are included, and together they request that cases be broken by *sex* and then by *fac*. It is requested that the table include the previously specified value labels for these break variables. This will identify the group to which each summary statistic applies. The summary subcommand requests a display (for each of the dependent variables) of the percentage of cases with a value greater than '1.5'. Since the three dependent variables involved have all been coded as 1 for 'true' and 2 for 'false', this cut-off will therefore identify those respondents who have reported that the statement does **not** apply i.e. for whom the statement is false. Also on this subcommand, the default summary label '**pcgt (1.5)**' has been replaced with the more informative '**% FALSE**'.

The above **REPORT** command (following the **SORT** command) produced the following output:

Subject sex	Home faculty	Often depressd	Sex thoughts often	Often anxious
Male	arts			
	% FALSE	66.67	16.67	33.33
	science			
	% FALSE	62.50	37.50	62.50
	soc. studies			
	% FALSE	37.50	37.50	37.50
Female	arts			
	% FALSE	25.00	66.67	16.67
	science			
	% FALSE	25.00	50.00	25.00
	soc. studies			
	% FALSE	41.67	66.67	25.00

14.3 Display format and layout

Additional **REPORT** subcommands allow the user to tailor the spacing, etc., of the display, and if the procedure is to be used repeatedly to produce complex breakdowns, it may be worth becoming familiar with the full range of subcommands. As an alternative, however, **REPORT** (or another procedure such as **MEANS**) could be used merely to produce the basic summary statistics of break groups, and the listing output from the procedure could then be edited with a word-processor or other editor. In this way the format and spacing could be changed, and additional titles and labels provided.

Such supplementary editing could be performed off-site on an ancillary microcomputer. One advantage of formatting with an editor is that the user is able to see on-screen the effects of adding spaces, titles, etc.

SPSS/PC+ and the word-processor

One of the great advantages of using SPSS/PC+ rather than other versions of SPSS is that files can easily be transferred between a word-processor and the data analysis system. Although use of the SPSS/PC+ editor program **REVIEW** has been encouraged in previous chapters, many other editors are also suitable for creating data files and data definition files (see Appendix B for general information on using other editor programs to create and edit SPSS/PC+ files). If **REVIEW** is used to create all of the files then the powerful SPSS/PC+ resource may be largely taken up by repetitive and routine work such as data entry.

Considerations of time and economy may therefore lead the user towards the use of an ancillary microcomputer for routine work. One other potential advantage of using a word-processor to create files, and one which might prove highly attractive to the novice user, is simply that of familiarity. Using a familiar system such as WordStar may enable the user to create and correct files, and to customize output with great ease.

There are clear advantages of becoming familiar with **REVIEW**, however, and for the efficient use of the SPSS/PC+ system it is advisable to use **REVIEW** as the within-session editor. When an SPSS/PC+ analysis session has ended, however, and the user has been returned to **DOS**, one or more of the files which have been automatically created by SPSS/PC+ can be copied. Thus the file containing the output for the session (saved by default as "SPSS.LIS") can be copied to a diskette on **drive A**. The following command could be used to accomplish this:

```
C> copy spss.lis a:
```

This would produce a copy of the listing file from the current directory on **drive C** (the hard disk) to the diskette on **drive A**. Using the above command, the copy would be saved with the same name (i.e. "SPSS.LIS").

An alternative name could be specified for the copy written to the diskette, e.g.

```
C> copy spss.lis a:gender1.lis
```

If the user has relied upon the default name for the listing file the file will be re-initialized the next time that SPSS/PC+ is run. This can be prevented, and a copy of the listing maintained on the hard disk, by making a second copy of the file on **drive C**:

```
C> copy spss.lis gender1.lis
```

or by renaming the listing file:

```
C> ren spss.lis gender1.lis
```

The copy of the listing file can now be used as a word-processing file, either on the same machine or on another computer to which the diskette has been transferred.

Incorporating SPSS/PC+ information into report texts

The SPSS/PC+ output file will be a copy of the output screen display (and printer output), although sometimes particular symbols (for example, those forming the axes of plots) may have changed somewhat. Corrections can be made with the word-processor (perhaps using a 'find and change string' facility).

Output page numbers, page divisions and certain SPSS/PC+ headings may need to be deleted, as well as error messages and the like. Some of the output will prove to be of little interest, and such redundant information can be deleted at this stage.

The information which is to be used within the report may be edited into a more suitable form. Line spacing can be changed, and the width of the output altered (if your editor program has a column edit facility this may prove especially useful). Additional titles and subtitles may be supplied, and existing text material may be underlined or emboldened for emphasis. Explanatory text may also be inserted.

The edited output file (often considerably reduced in size) can then be saved, either as a single file or in the form of a number of small aptly titled files. When the edited file has been saved, the original output file will be available only as a backup file (usually with the file extension **.BAK**). Any future editing of the new file will lead to a loss of this original file, and it is therefore good practice to rename the backup file and thus to preserve a copy of the original output. If this is not done then some of the information from the file may be lost altogether. The user may later identify a need for some information which has not been included within the revised file. If a copy of the original information has been retained, however, this will not present a problem.

During the phase of report writing, various fragments of the output from the SPSS/PC+ analyses, now edited, can be incorporated into the text of the report as it is being written. Further editing might be necessary at this stage, and care should be taken that tables are not split awkwardly across text pages. Some reports will consist largely of the output material itself, with little additional commentary, while other reports will consist mostly of text material with just the occasional inclusion of tables and graphs.

15   Joining Files

15.1 Adding new data

Sometimes, after a file has been created, new data are obtained and the user wishes to add them to the original file. New **variables** or new **cases** may need to be added.

New variables: Suppose that a file contains the results of a political attitude survey taken before an election, and that information has subsequently been collected from many of the same subjects **after** the election. If the user wishes to analyze changes and to examine relationships between some of the original and some of the new variables, the relevant variables must all be contained within a single file. A composite file therefore needs to be created, containing both the old and the new **variables**.

New cases: Another situation might arise. Suppose that a large number of questionnaires have been sent out in a postal survey. After the first batch of questionnaires has been returned a data file may be created, but for some time afterwards further returns may continue to arrive. Eventually a final summary and analysis of the results will be needed for all of the cases together. A composite file will need to be created, containing both the old and the new **cases**.

One simple and direct way of expanding an existing file to add new cases or new variables would be to return to the original data file and to insert the additional information with an editor program (such as **REVIEW**), amending the data definition file as appropriate.

Alternatively, the new data (i.e. the new cases, or new variables for the original cases) could be entered into a second file and a **JOIN** command then used to create a new composite file. This would include both the information from the original file and that from the second file.

The **JOIN ADD** command is used to add new **cases**: the **JOIN MATCH** command is used to add new **variables**.

Many of the same rules apply to the **JOIN MATCH** and **JOIN ADD** commands:

- up to five input files can be joined to form a single composite file
- the files to be joined must be **system** files (except that one can be the

current active file – the active file is specified in the **JOIN** command by an asterisk ⋆)
- unless a special order subcommand (By) is in effect, the order of the cases or variables in the composite file is determined by the order in which the input files are named on the command

Several **subcommands** can also be used within **JOIN** commands:

- the user can **Drop** specified variables from the files to be joined
- the user can **Keep** specified variables from the files to be joined
- the user can **Rename** variables before entering them into the new composite file
- the user can issue a **Map** subcommand to identify the source of variables in the composite file (and the effects of any **Rename** instructions)

The **LIST** command can be used to list the data in the new file, and the **SAVE** command can be used to save the contents of the new file on disk.

15.2 The JOIN MATCH command

The **JOIN MATCH** command is used to create a new file containing the **variables** from two files (for overlapping cases). Suppose that the simple data definition file:

```
DATA LIST FILE= "file1.dat" / id 1-3 var001 to var010 4-13.
```

refers to a data file ("file1.dat") which includes 10 cases and contains the following data:

```
0011111111111
0021111111111
0031111111111
0041111111111
0051111111111
0061111111111
0071111111111
0081111111111
0091111111111
0101111111111
```

To **JOIN** the data from this file with those from another file first create a system file (e.g. "file1.sys") containing the data. To create such a file from the data definition file ("file1.def"), we would respond to the SPSS/PC+ prompt with 'include file="file1.def" ' and then create a system file by using the **SAVE** command. Thus:

```
SPSS/PC: include file= "file1.def".
SPSS/PC: save outfile= "file1.sys".
```

Imagine that further data has been collected for the same subjects.
The new data definition file ("file2.def") is:

```
DATA LIST FILE= "file2.dat" / id 1-3 var011 to var020 4-13.
```

and the new data file ("file2.dat") contains the following data:

```
0022222222222
0022222222222
0032222222222
0042222222222
0052222222222
0062222222222
0072222222222
0082222222222
0092222222222
0102222222222
```

Again, we would create a **system file** ("file2.sys") from the data definition file. With versions of both of the relevant files as system files we can now create a composite file with a **JOIN** command:

```
SPSS/PC: JOIN MATCH FILE= "file1.sys"
       : /FILE= "file2.sys".
SPSS/PC: LIST.
```

(If "file2.def" had been used to create the current active file then the second subcommand above would be replaced with '/FILE = $\star$.').

Although the variable *id* is included in both of the original files, it will occur only once in the composite file. This is because the variable has the same **name** in both of the original files. If different names had been used, the composite file would include the subject identifier twice, under different names.

The order of the filenames on the **JOIN MATCH** command ("file1.sys" followed by "file2.sys") determines the order in which the variables will appear on the composite file.

The above command would produce an active file composed of material from both of the source files. The **LIST** command would produce the following display:

ID	V A R 0 0 1	V A R 0 0 2	V A R 0 0 3	V A R 0 0 4	V A R 0 0 5	V A R 0 0 6	V A R 0 0 7	V A R 0 0 8	V A R 0 0 9	V A R 0 1 0	V A R 0 1 1	V A R 0 1 2	V A R 0 1 3	V A R 0 1 4	V A R 0 1 5	V A R 0 1 6	V A R 0 1 7	V A R 0 1 8	V A R 0 1 9	V A R 0 2 0
1	1	1	1	1	1	1	1	1	1	1	2	2	2	2	2	2	2	2	2	2
2	1	1	1	1	1	1	1	1	1	1	2	2	2	2	2	2	2	2	2	2
3	1	1	1	1	1	1	1	1	1	1	2	2	2	2	2	2	2	2	2	2
4	1	1	1	1	1	1	1	1	1	1	2	2	2	2	2	2	2	2	2	2
5	1	1	1	1	1	1	1	1	1	1	2	2	2	2	2	2	2	2	2	2
6	1	1	1	1	1	1	1	1	1	1	2	2	2	2	2	2	2	2	2	2
7	1	1	1	1	1	1	1	1	1	1	2	2	2	2	2	2	2	2	2	2
8	1	1	1	1	1	1	1	1	1	1	2	2	2	2	2	2	2	2	2	2
9	1	1	1	1	1	1	1	1	1	1	2	2	2	2	2	2	2	2	2	2
10	1	1	1	1	1	1	1	1	1	1	2	2	2	2	2	2	2	2	2	2

The composite file has now become the active file. It can be **SAVED** as a system file by using a command such as:

```
SPSS/PC: SAVE OUTFILE= "compfile.sys".
```

The two input files which were joined in the above example ("file1.sys" and "file2.sys") are said to be **parallel** because each contains data for the same cases in exactly the same order. The second file contains extra **variables** for **every** case in the first file. There are no missing cases and no extra cases.

In real studies such a situation would be rare. Commonly, follow-up data is obtained from only **some** of the original cases. Suppose that the second data file contained data from only 7 of the original 10 cases, respondents 4, 6 and 9 having failed to reply to a follow-up questionnaire. Unless the **JOIN MATCH** command above were modified, the system would assign the data from the second phase of the study to subjects 1 to 7. The system would treat the remaining subjects (8 through 10) as if they had not provided follow-up data (it would assign system missing values to these **cases** for each of the follow-up variables). Clearly this would provide a serious mismatching which would produce major errors in subsequent analyses.

Such mismatching can be prevented by providing instructions on how the data in the second file should be assigned to cases from the first file. The most efficient way of doing this is to include a subject *id* variable in **each** of the files to be joined, and then to instruct the program to match **BY** *id*. The **BY** subcommand must follow after all other file specifications. Thus:

```
SPSS/PC: JOIN MATCH FILE= "file1.sys"
       : /FILE= "file2.sys"
       : /BY id.
SPSS/PC: LIST.
```

This would produce a composite file containing some system-missing values, and these would be indicated by a period (.) in a subsequent **LIST** display. Thus with no second-phase data available for subjects 4, 6 and 9, the **LIST** command would produce the following output:

	V A R 0 1	V A R 0 2	V A R 0 3	V A R 0 4	V A R 0 5	V A R 0 6	V A R 0 7	V A R 0 8	V A R 0 9	V A R 1 0	V A R 1 1	V A R 1 2	V A R 1 3	V A R 1 4	V A R 1 5	V A R 1 6	V A R 1 7	V A R 1 8	V A R 1 9	V A R 2 0
ID																				
1	1	1	1	1	1	1	1	1	1	1	2	2	2	2	2	2	2	2	2	2
2	1	1	1	1	1	1	1	1	1	1	2	2	2	2	2	2	2	2	2	2
3	1	1	1	1	1	1	1	1	1	1	2	2	2	2	2	2	2	2	2	2
4	1	1	1	1	1	1	1	1	1	1	.	.	.	.	.	.	.	.	.	.
5	1	1	1	1	1	1	1	1	1	1	2	2	2	2	2	2	2	2	2	2
6	1	1	1	1	1	1	1	1	1	1	.	.	.	.	.	.	.	.	.	.
7	1	1	1	1	1	1	1	1	1	1	2	2	2	2	2	2	2	2	2	2
8	1	1	1	1	1	1	1	1	1	1	2	2	2	2	2	2	2	2	2	2
9	1	1	1	1	1	1	1	1	1	1	.	.	.	.	.	.	.	.	.	.
10	1	1	1	1	1	1	1	1	1	1	2	2	2	2	2	2	2	2	2	2

Each of the input files to be included in the composite file must already be sorted in the ascending order of any variable used (by being specified after the **BY** keyword) to bring about a match. Complex matching procedures can be specified (with up to 10 **BY** keywords used within a single command), but the simplest way of arranging a match is to include the *id* variable for each respondent on each of the input files. If the data are then entered (or re-sorted) in ascending *id* order, the files can be matched using the **BY** subcommand ('BY *id*').

The Drop and Keep subcommands

Variables can be dropped from the composite file by specifying **Drop**, followed by a variable list, after the relevant input file name. Thus:

```
SPSS/PC: JOIN MATCH FILE= "file1.sys"
       : /DROP=var004
       : /FILE= "file2.sys"
       : /DROP= var011 var012 var013 var018
       : /BY id.
SPSS/PC: LIST.
```

If only a few of the variables from a particular file are to be entered into the composite file it may be more efficient to specify which ones should be kept. In this case a **Keep** subcommand would be used. This should specify, after the relevant input file name, the list of the variables to be included in the composite file.

N.B. The TO convention is not permitted when variables are listed on Drop or Keep subcommands within a JOIN command.

The Rename subcommand

Variables can be renamed before they enter the composite by specifying **Rename** and listing the changes to be made in the following manner:

```
SPSS/PC: JOIN MATCH FILE= "file1.sys"
       : /RENAME (var001 var003 var005 = att1 att3 att5)
       : /DROP=var004
       : /FILE= "file2.sys"
       : /RENAME (var014 = item3)
       : /DROP=var011 var012 var013 var018
       : /BY id.
SPSS/PC: LIST.
```

Notice that in the **Rename** subcommand either a single variable or a number of variables may be renamed within a pair of brackets with a = sign between the old and the new variables, thus:

```
(old1 = new1)    or    (old1 old2 = new1 new2)
```

N.B. The TO convention is not permitted on the Rename subcommand of the JOIN command.

The above command, with the **Drop** and **Rename** subcommands as specified, produced the following listing:

```
         V           V   V   V   V   V        V   V   V   V   V
         A           A   A   A   A   A   I    A   A   A   A   A
      A  R  A  A     R   R   R   R   R   T    R   R   R   R   R
      T  0  T  T     0   0   0   0   0   E    0   0   0   0   0
      T  0  T  T     0   0   0   0   1   M    1   1   1   1   2
  ID  1  2  3  5     6   7   8   9   0   3    5   6   7   9   0

   1  1  1  1  1     1   1   1   1   1   2    2   2   2   2   2
   2  1  1  1  1     1   1   1   1   1   2    2   2   2   2   2
   3  1  1  1  1     1   1   1   1   1   2    2   2   2   2   2
   4  1  1  1  1     1   1   1   1   1   .    .   .   .   .   .
   5  1  1  1  1     1   1   1   1   1   2    2   2   2   2   2
   6  1  1  1  1     1   1   1   1   1   .    .   .   .   .   .
   7  1  1  1  1     1   1   1   1   1   2    2   2   2   2   2
   8  1  1  1  1     1   1   1   1   1   2    2   2   2   2   2
   9  1  1  1  1     1   1   1   1   1   .    .   .   .   .   .
  10  1  1  1  1     1   1   1   1   1   2    2   2   2   2   2
```

Here, the *id* variable and the next nine variables have come from the input file "file1.sys". One of the variables from this source file has been dropped and three have been renamed. The remaining six variables come from "file2.sys". Four of the original variables in this file have been **dropped** and one (*id*) is not featured because it is a duplication of the *id* variable from the first named file. One of the variables has been renamed. The file "file2.sys" contains only seven cases, and system missing values have been assigned where **cases** are missing. The files have been matched using the **BY** keyword.

The contents of the composite file

If variable labels, value labels and missing values are specified on any of the input files then the composite file will retain any of the information which is relevant to the data it contains. If two or more files contain **different** definitional information regarding particular data, then the composite file will include the version which was **first** encountered in the specified sequence of input files.

Similarly, if the same variable **name** is encountered in more than one of the input files, the value entered into the composite file will be the **first** value encountered (even if this is a missing value).

15.3 The JOIN ADD command

Suppose that we have created a file of the initial returns from a postal survey and we now wish to add the data from late returns. We thus have new **cases** with the same variables as the original cases. To combine the data from the early and the late returns we can create a new file with **JOIN ADD**.

As with **JOIN MATCH**, the input files to be combined into a composite file must be system files (except that one can be the active file). The simplest form of the **JOIN ADD** command is:

```
SPSS/PC: JOIN ADD FILE= "data1.sys"
       : /FILE= "data2.sys".
```

This would create a new (active) file which would contain the cases from "data1.sys" followed by the cases from "data2.sys".

Thus if the first input file contains 100 cases, and the second input file contains 56 cases, the composite file will contain 156 cases. Each case on the composite file will contain all of the variables included on **any** of the input files (system missing values will be provided if a file does not contain a variable which is included in one of the other input files).

Suppose that 500 questionnaires, pre-coded 001 to 500, have been mailed. When some of the completed questionnaires have been returned an initial data file may be created. Rather than sorting the questionnaires physically into the code order we could simply work through the pile of returns, entering first the code number (as the variable *id*), followed by the other data. The first data file might contain data from 200 questionnaires. The order of the cases in that file, in terms of the *id* numbers, will be haphazard or random.

After a further month we may decide that the data collection is now at an end. Perhaps another 150 questionnaires will have been received, making 350 (or a 70 per cent return) altogether. Data from the second batch can be entered into a second file, again in haphazard *id* order.

If the two datasets are to be added together (using **JOIN ADD**), each will first need to be defined by a data definition file. The same variable names should be used for each of the parallel files. System file versions of each of the files need to be created (except for one which may be the active file), and the files can then be joined by a **JOIN ADD** command, e.g.

```
SPSS/PC: JOIN ADD FILE= "data1.sys"
       : /FILE= "data2.sys".
```

The composite file could then be re-ordered with a **SORT** command so that cases appear in their original *id* order:

```
SPSS/PC: SORT BY id.
```

The re-ordered active file could then be saved:

```
SPSS/PC: SAVE OUTFILE= "combined.sys".
```

As with **JOIN MATCH**, up to five files can be joined with **JOIN ADD**.

The subcommands **Rename**, **Drop** and **Keep** can be used to select and name the **variables** within the composite file (**cases** cannot be dropped). **Rename**, **Keep** and **Drop** subcommands apply only to the file named immediately before, and will therefore need to be repeated if the same variables are to renamed, kept or dropped from different source files. Thus:

```
SPSS/PC: JOIN ADD FILE= "data1.sys"
       : /RENAME= (var001 var003 var005 = item1 item3 item5)
       : /DROP= var002 var004 var006 var007 var008 var009
       : /FILE= "data2.sys"
       : /RENAME= (var001 var003 var005 = item1 item3 item5)
       : /DROP= var002 var004 var006 var007 var008 var009
       : /BY id /map.
```

Notice that each of the variables to be dropped is specifically named. The **TO** convention is not permitted on a **Rename**, **Keep** or **Drop** subcommand of a **JOIN** command.

The **Map** subcommand can be used at any stage within the command to list the variables on the active file together with their source. If **MAP** is included as the final subcommand it will show the final state of the active file (with the heading 'RESULT'). The **Map** request within the above command produced the following output:

```
RESULT          DATA1.SYS       DATA2.SYS
------------    ------------    ------------

ID              ID              ID
ITEM1           VAR001          VAR001
ITEM3           VAR003          VAR003
ITEM5           VAR005          VAR005
VAR010          VAR010          VAR010
```

16 Additional Topics

This chapter concerns three topics:

- some further ways in which data can be entered into SPSS/PC+ files
- how **portable files** may be created and transported between the SPSS/ PC+ microcomputer and other computers (including mainframe computers running SPSS[X])
- a brief mention of some commands from the SPSS/PC+ base system which have not been included in this presentation, and some of the facilities offered by the various optional enhancements that are available

16.1 New data options

By this time the *SPSS/PC+ Manual* is likely to have become an indispensable aid, and you should now have become familiar enough with the SPSS/PC+ system to be able to understand most of the material contained within it.

This book will have provided you with enough information for an initial exploration of a dataset, and you should now have a good overview of the nature of the SPSS/PC+ system. The presentation has avoided many of the ifs and buts which would have been necessary to provide a truly comprehensive account. By now, however, you should be skilled enough, and confident enough, to explore further ways of using the system.

In this section we will consider some alternative ways in which data may be entered. Thus, as well as the numerical data which we have focused upon, SPSS/PC+ also allows alphanumeric variables (the names of people, or of countries, for example) to be represented by **strings**. The strategy we have followed in assigning particular columns in the dataset to particular variables (fixed format) is actually one of two ways in which data can be arranged (the other way is free format). And whereas we have been assuming that all of our data is integer, SPSS/PC+ provides a facility for entering decimal data in such a way that the decimal point is assumed.

Strings

Throughout this book, all of the data have been represented in the form of numbers whether the values have been truly numerical or categorical. But SPSS/PC+ can also deal with alphanumeric data, for example, the names and addresses of respondents.

Thus a data file ("a.dat") could look like this:

```
001   Mary     Dobson   21433123145645621
002   John     Croft    12433123145645621
003   Robert   Crystal  11433123145645621
004   Susan    Lewin    22433123145645621
```

A data list ("a.def") for the above file might be:

```
DATA LIST FILE = "a.dat"/ id 1-3 forename 6-12 (a) surname 16-23 (a)
sex 24 agegroup 25 var001 to var013 26-38.
```

The (A) specification which follows the column allocation for the variables *forename* and *surname* indicates that these two variables are strings.

Strings which are eight characters long or less are referred to as short strings. Other strings (which may be up to 255 characters in length) are referred to as long strings. Each eight-character portion of a long string counts as one variable towards the system limit of 200 variables (thus a nine-character string would count as two variables, a 20-character string as three variables, etc.). Certain manipulations are possible with short strings but not with long strings – for example, only short strings can be used in transformation commands (e.g. **COMPUTE** and **RECODE** commands).

When a string is specified in a **command** it must be enclosed within quotation marks, and cannot be split across two lines.

A **short** string variable can be created with an **IF** command. In the following example:

```
SPSS/PC: if (sex eq 1) sexalph = "  male".
SPSS/PC: if (sex eq 2) sexalph = "female".
```

male cases would be assigned the value ' male' for the new alphanumeric variable *sexalph*, and the value 'female' would be assigned to this variable for female subjects.

If short strings are included in transformation commands, the specification must be **precise**. Leading or trailing blanks must be included, and lower- and upper-case characters are **not** treated as equivalent. Strings cannot be compared to numbers, or to numerical variables. The following set of commands was used to create a new variable *group* from an original numerical variable *agegroup* and the newly created short string variable *sexalph*:

```
SPSS/PC:  if (sexalph eq "  male" and agegroup eq 1) group = "BOY".
SPSS/PC:  if (sexalph eq "  male" and agegroup eq 2) group = "MAN".
SPSS/PC:  if (sexalph eq "female" and agegroup eq 1) group = "GIRL".
SPSS/PC:  if (sexalph eq "female" and agegroup eq 2) group = "WOMAN".
```

Used with the data from the file "a.dat", and after *sexalph* had been computed, this set of commands created the new short string alphanumeric variable *group*. The following frequencies display was then produced in response to the command **FREQUENCIES group.**:

```
GROUP
                                               Valid     Cum
    Value Label     Value  Frequency  Percent  Percent  Percent

                      BOY      1        25.0     25.0     25.0
                      MAN      1        25.0     25.0     50.0
                     GIRL      1        25.0     25.0     75.0
                    WOMAN      1        25.0     25.0    100.0
                            -------   -------  -------
                    TOTAL      4       100.0    100.0
```

A **SORT** command specifying a string can order cases alphabetically. In such a sort, numbers will come before letters and upper-case letters (A–Z) will come before lower-case (a–z). Such alphabetic sorting can either be ascending (A–Z, etc.) or descending (Z–A, etc.), e.g.

```
SPSS/PC:  sort by forename.
SPSS/PC:  sort by surname (D).
```

On the **MISSING VALUE** command, the specification for a missing string variable must be specified within quotation marks. Thus:

```
MISSING VALUES var001 to var056 (9) forename "       ".
```

Freefield formatting

In previous chapters, in describing how data should be set out within the data file, specific columns were always allocated to particular variables, and it was stressed that any variable must occupy exactly the same column(s) for all cases.

In doing this, however, we were restricting the data arrangement to the default fixed format. There is an alternative format arrangement known as free format. In free format, variables are separated by one or more blanks (and/or a comma). Thus the system would automatically recognize the dataset:

```
001 2 34 456 45 5 5 6 4 54 64 23 13 678 78
```

as containing 15 variables. When free format is used, data do not have to be aligned by columns, so that the following dataset would be permissible:

```
001 2 34 456 45 5 5 6 4 54 64 23 13 678 78
002 12 3 46 3 3 2 1 4 46 34 54 12 324 67
003 3 23 324 34 4 5 3 3 41 32 23 9 345 65
004 8 2 23 1 1 2 2 3 47 45 13 23 67 54
```

Each case here contains 15 variables, including *id*. The variables must be in the same order for each case. Although the leading zeros (00 in 001, for example) have been retained in this illustration (and **could** be included in a real free format dataset), they serve no useful purpose in a free format situation. Identification numbers, therefore, could be entered as 1, 32, 465, etc. without the leading zeros used in fixed format to control the alignment.

If free format is to be used, the keyword **FREE** must be specified after **DATA LIST**, e.g.

```
SPSS/PC: DATA LIST FREE / var001 to var015.
```

The data from more than one case can be entered on a single record, so that the data read as a result of the following command:

```
SPSS/PC: DATA LIST FREE / var001 to var002.
```

might all be contained on the following single record:

```
34 45 12 8 34 56 54 52 45 9 56 56
```

(this would represent the data for six cases with two variables per case).

If the **DATA LIST** command refers to a file, then the command takes the form:

```
SPSS/PC: DATA LIST FILE= "genfree.dat" FREE / var001 to var015.
```

By default, all data are assumed to be numeric, but string variables can be used with a freefield format. String variables must be identified by **(A)** following the variable name. By default, strings are assumed to be eight characters in length. The format can be changed by specifying a width within the **(A)** brackets:

```
SPSS/PC: DATA LIST FREE file="mand.dat" /var001 surname (A15) var002.
```

The data file ("mand.dat") might contain the following information:

```
321 "Mandeville"  2
7352 "Nolan" 12
823 "Postlethwaite" 23
45  "Nye" 14
```

Here the names are shorter than the 15 columns allocated. The system would allocate padding blanks to the right, however, so that the strings would be treated as:

```
"Mandeville     "
"Nolan          "
"Postlethwaite  "
"Nye            "
```

If any future reference was made to these strings the user would have to include the extra blanks.

There is little scope for using long strings in any manipulation or analysis, although they can be **LIST**ed and used as **SORT** variables. For example:

```
SPSS/PC: sort by surname.
```

If you attempt to use a long string variable in a procedure such as **FRE-QUENCIES** or **CROSSTABS** you will receive a warning '**invalid long string**', but the system will carry out the procedure with the variable truncated to eight characters. This generally presents no problem, although when the name 'Postlethorpe' was added into the above file, the **FREQUENCIES** output did report the presence of two 'Postleth's in the dataset.

Non-integer data

All of the data used in this book have been integer (i.e. there have been no decimal points within the dataset). Sometimes, however (especially with data from experimental studies rather than surveys) some of the data will be decimal data (e.g. 345.566, 23.67, 40.002). There is no problem in including such numbers within a dataset for use by SPSS/PC+ system, although they will generally represent values for dependent rather than independent variables.

Suppose we have the dataset shown in Table 16.1.

Table 16.1

var001	var002	var003	var004	var005
23	−345.3	14.004	34	.003
35	45.7	12.234	45	.123
5	142.5	9.030	8	.045
78	−45.6	18.675	27	.067

This set of numbers could be written in a data file (say, "dec.dat") in precisely this form, with the decimal point included. The following **DATA LIST** command might be used to define the variables:

```
SPSS/PC: DATA LIST file = 'dec.dat' / var001 8-9 var002 14-19
       :   var003 24-29 var004 34-35  var005 43-46.
```

Fixed format is the default format and need not be specified by the user. The columns allocated to particular variables should make allowance for the decimal point and for the minus sign (−) (each of which would occupy one column).

The above dataset could also be defined by a **DATA LIST** command which specifies a **FREE** format:

```
SPSS/PC: DATA LIST FREE / var001 to var005.
```

If a free format is to be used, the dataset could also be written:

```
23 -345.3 14.004 34 .003
35 45.7 12.234 45 .123
5 142.5 9.030 8 .045
78 -45.6 18.675 27 .067
```

For **fixed format data lists**, the decimal points may be omitted from the dataset and information added to the DATA LIST command to specify how many decimal places should be **assumed** for each variable. The name of a variable for which a decimal point is to be assumed must be followed by the column specification (as usual) and by a number in brackets specifying the decimal places to be assumed to the **right** of the decimal point. This is illustrated in the following example:

```
SPSS/PC: DATA LIST file = "point.dat" / id 1-3 var001 4-5
       : var002 6-7 (1) var003 8-11 (3) var004 12-14 (2)
       : var005 15-16 (2).
```

If the data file "point.dat" contained the following numbers:

```
                    1111111
Columns         1234567890123456
                ================
                9645873443553285
                8568356326443228
                3568932456984675
                2345692345692356
```

it would represent the data for each of the variables in this way:

Variables	id	v1	v2	v3	v4	v5
	===	==	==	====	===	==
	964	58	73	4435	532	85
	856	83	56	3264	432	28
	356	89	32	4569	846	75
	234	56	92	3456	923	56

The system, however, would treat the set of numbers as shown below (following the specifications regarding implied decimal points):

Decimal point specification	(0)	(0)	1	3	2	2
Variables	id	v1	v2	v3	v4	v5
	===	==	===	====	====	=====
	964	58	7.3	4.435	5.32	.85
	856	83	5.6	3.264	4.32	.28
	356	89	3.2	4.569	8.46	.75
	234	56	9.2	3.456	9.23	.56

Data referred to on **free format data lists** must include the decimal point. Instructions cannot be given for the system to **assume** a decimal point. The following example shows the contents of a data file ("frepoint.dat") which is to be read by a free format data list:

```
3.4   23.455   47.567 5.4 2.44
.76 12.228 23.444 23 44.4
3 78.123     78.993 43 15.4
83 456.66   23.444 78 56.7
43 838.38 56.454 32 523
```

If an additional command, **FORMAT**, is used, the decimal nature of the data can be tidied for presentation in later displays. The **FORMAT** command uses the **F-convention** and is of the form **F4.2**. This example would request that the variable associated with this format specification should be displayed (and written) with four number characters, two of which would be to the **right** of the decimal point. If more decimal places occur in the actual data, then when it is displayed, the value of the item is rounded to conform to the specified decimal format. Thus the following commands (referring to the data file "frepoint.dat" above):

```
SPSS/PC: data list free file = "frepoint.dat"/ var001 to var005.
SPSS/PC: formats var001 (F2.0) var002 (f5.1) var003 (f5.3)
       : var004 (f2.0) var005 (f3.1).
```

produced the following output from a **LIST** command:

VAR001	VAR002	VAR003	VAR004	VAR005
3	23.5	47.567	5	2.4
1	12.2	23.444	23	44.4
3	78.1	78.993	43	15.4
83	456.7	23.444	78	56.7
43	838.4	56.454	32	523.0

Data cleaning

It has been assumed that the utmost care has been taken in transferring data from a data sheet, or from the questionnaires, to the computer, and that mistakes have not been made. This assumption may be over-optimistic, however, and the following procedure can be used to detect **some** errors which may have occurred at the data entry stage.

If a variable can take, say, the three valid values 1, 2 and 3, and if that variable has been assigned the missing value 9, then clearly any case which has a value between 4 and 8, or 0, for that variable has been mis-coded, or a mistake has been made in entering the data.

Such errors can be detected by using the **FREQUENCIES** procedure. When you have created a data file and written the relevant data definition file (and having **SET** the printer **off**) you can issue the command:

```
SPSS/PC: frequencies all.
```

As the frequency table for each variable is displayed, consult the codebook to determine whether the system is reporting any occurrences of values other than valid and missing values. Make a note of any variables which have such an occurrence.

The next step is to identify which case(s) include an illegal value for any of the variables. Suppose that a variable *triad* has valid values 1, 2 and 3, but that a **FREQUENCIES** analysis has revealed that there is one case with the value 4 for *triad*. Issue the commands:

```
SPSS/PC: process if (triad eq 4).
SPSS/PC: crosstabs id by triad.
```

and you will be presented with a screen display that will identify which case has the value 4 for *triad*. Make a note of this so that after detecting all such errors you can consult the original data source and, having determined which column of the dataset contains the error, you can correct the data file.

No similar procedure will enable you to identify cases for which a value within the valid range has been substituted for the true value, and no cleaning operation such as that described can guarantee that errors do not remain within the dataset. It is very important, therefore, to ensure total accuracy when creating the data file, and to check a printed copy of the file against the original data source.

16.2 Transporting files between computers

Portable files

Files can be transferred, or transported, between different computers if they are written in a portable form. **Portable files** are created using the command **EXPORT**, and a copy can be sent to another machine via a special computer line, or a normal telephone line, using a **file-transfer program**. The file-transfer program usually employed to transport SPSS/PC+ files is called **Kermit**. With **Kermit** in action on both computers and a portable file created on one of them, that machine can send a copy of the file to the other computer. The transferred file can then be read using an **IMPORT** command. Portable files are not in the standard text file format and, generally, should not be edited. When an **IMPORTED** file has been made active it can be **SAVED** as a system file.

The EXPORT command
EXPORT is used to create a portable file from the current active file. The

portable file is written to the current directory and can later be transported (using **Kermit**) from SPSS/PC+ to another system (such as **SPSS^X** on a mainframe computer or **SPSS/PC** on another microcomputer). If a portable file is to be uploaded to another computer then suitable (and compatible) file-transfer programs must be loaded on **both** machines. **EXPORT** can also be used on another machine (for example, a mainframe computer running **SPSS^X**) to create a portable file to be exported (via **Kermit**) to the SPSS/PC+ microcomputer.

EXPORT writes the current active file, including the data itself, the data dictionary, and certain variables assigned by the system. Although portable files are written in alphanumeric (ASCII) format they are not easily editable.

The simplest **EXPORT** command includes the command word followed by **outfile=** and the name of the portable file to be created, for example

```
EXPORT OUTFILE= "gender.por".
```

In this case the entire contents of the active file would be placed in a new export file named "gender.por". The variables would retain their original names and values together with the original variable labels, value labels and missing value flags. Changes to the active file made as a result of any **COMPUTE, RECODE, SELECT** and other commands issued during the session will remain in effect when a portable file is created.

It is possible to write a portable file that includes only **some** of the variables in the active file. This can be done either by specifying the variables to be retained (using a **Keep** subcommand) or those to be dropped (using a **Drop** subcommand). Thus:

```
SPSS/PC: EXPORT OUTFILE= "gender.por" / KEEP= id to marit.
```

or

```
SPSS/PC: EXPORT OUTFILE= "gender.por" / DROP= age mar01 to e12.
```

The portable file created by the second of these examples would include all of the variables on the active file **except** the variables listed.

Judicious use of a **DROP** or a **KEEP** command may be essential when creating a portable file on a mainframe for later transportation to SPSS/PC+. A mainframe has far more capacity and power than a microcomputer, and if the portable files produced from **SPSS^X** are too large then the microcomputer may not have sufficient memory capacity to deal with them. A portable file created by SPSS/PC+, or a portable file to be read by SPSS/PC+, cannot contain more than 200 variables.

Variables in the active file may be renamed as the portable file is created. Thus:

```
SPSS/PC: EXPORT OUTFILE= "gender.por"
       : / rename= (st01 st03 = depress anxious).
```

Here *st01* would be renamed *depress*, and *st03* would be renamed *anxious*.

If changes are made to the file (with **Drop**, **Keep** or **Rename**) it is useful to keep within the portable file a listing of the changes. This will help the user to keep track of how the portable file relates to the original file from which it was created. Such a listing is requested by including **Map** as a final subcommand:

```
SPSS/PC: EXPORT OUTFILE= "gender.por"
       : / rename= (st01 st03 = depress anxious)
       : / drop= dat01 to dat08 / map.
```

Changes made in creating a portable file do not affect the active file.

Having created an exportable file the user will probably want to send this file to another computer using **Kermit**. Before considering how this is done, however, we will consider the use of the **IMPORT** command. This is used to read a portable file into the SPSS/PC+ system and is similar to **EXPORT** in many ways.

The IMPORT command

IMPORT is a 'data read' command (i.e. a data definition command rather than a procedure) which reads information from a portable file. When a subsequent procedure command is issued the active file is created. The portable file must be in the current directory although it can be on another disk-drive (for example, the diskette drive, **drive A**, of the micro-computer). Typically the portable file will have been downloaded, via **Kermit**, from a mainframe computer running **SPSS**[X].

The simplest **IMPORT** command includes the command word followed by **file=** and the name of the portable file to be read. Thus, e.g.

```
SPSS/PC:  IMPORT FILE= "mainfram.por".
```

In this case the entire contents of the portable file would be read. To read only some of the variables within the portable file a **Drop** subcommand or a **Keep** subcommand would be used, specifying those variables to be dropped or kept. Thus:

```
SPSS/PC: IMPORT FILE= "mainfram.por" / DROP= item45 to item54.
```

If **Drop** is used then the variables will appear in the active file in the same order as in the portable file. If **Keep** is used, however, the order in which the variables are named will determine their order within the active file.

Variables may also be renamed as they are read from the portable file, using a **Rename** subcommand. Thus:

```
SPSS/PC: IMPORT FILE= "mainfram.por"
       : / rename= (item1 to item44 = var001 to var044).
```

This would simply rename all 44 of the *itemX* variables as the equivalent *varXXX* variables. Thus *item1* would become *var001*, *item41* would become *var041*, etc. The **TO** convention is permitted in **IMPORT** and **EXPORT** commands (you may remember that it was not permitted in **JOIN** commands).

As with **EXPORT**, a listing of any changes made during the **IMPORT** (with **Drop**, **Keep**, or **Rename**) can be included in the file. Once again, this is done using **Map** as the final subcommand: e.g.

```
IMPORT FILE= "mainfram.por" / rename= (item45 = total)
/ drop=item20 to item23 / map.
```

Having discussed how an exportable portable file can be created with **EXPORT** and how an importable portable file can be read with **IMPORT** we now need to consider how such files may be transferred from one computer to another.

Using Kermit to transfer portable files

Two computers may be linked by a special line, or one computer may communicate with another via a telephone link using a piece of apparatus known as a **modem**. The host computer (which must also be connected to a modem) is then dialled. With the appropriate commands the user introduces him/herself to the host machine (using log-on instructions, special passwords and the like) and information can then be transferred between the two machines. Commands issued from the keyboard of the home (or resident) computer can be made to operate programs in the remote or host machine.

When transferring SPSS files, the **Kermit** program is generally used. This allows files created by **SPSS**[X] on a mainframe computer to be transported to a SPSS/PC+ microcomputer (and vice-versa). **Kermit** is error correcting, so that after a successful file transfer the user can be sure that the copy is a true replica of the original.

Details on using **Kermit** are provided in Section F of the *SPSS/PC+ Manual*. To access the mainframe you will, of course, need to be registered as a user. Certain parameters of the communication line will need to be appropriately set (these include the baud rate – the signalling speed – and the parity – an error-checking strategy). Because the setting of certain parameters is crucial to the success of file transfer, your first attempt at using **Kermit** may not prove entirely straightforward. If other users working with your system have experience in the successful transportation of files to and from the other computer then they may be willing to demonstrate the process to you and to explain how the particular link functions. Many different versions of **Kermit** are available and they are variously implemented on **IBM**, **Honeywell**, **VAX**, **ICL** and many other mainframe computers. Compatible versions of **Kermit** need to be available on the mainframe and the microcomputer.

To operate file transfers from the SPSS/PC+ microcomputer, the **Kermit** program is run (through **DOS**) and a connect command issued. The appropriate telephone number is dialled to connect to the mainframe computer and, after an acknowledgement from the host computer has been received, the user logs on to the host.

At this stage the resident machine acts as a terminal of the mainframe. The host computer can be requested to run its own **Kermit** program to send a portable file, e.g.

```
SEND mainfram.por
```

The user then escapes back to the microcomputer (for example, by pressing <Control]> and then <C> – a screen display will usually provide information on the appropriate escape sequence).

The microcomputer **Kermit** program is then instructed to receive a named portable file which has previously been created on the mainframe by using the **EXPORT** procedure, e.g.

```
RECEIVE mainfram.por
```

Transfer should now occur, and the requested portable file should be written to the current directory of the microcomputer. The user now regains connection with the host computer in order to log off and end the connection to the mainframe.

When the portable file is **IMPORTED**, e.g.

```
SPSS/PC: import file= "mainfram.por".
```

it can become the active microcomputer file, ready for transformation and analysis. And once the information from the portable file has been made active it can be **SAVED** as a system file.

The process of uploading a portable file from a microcomputer to a mainframe is very similar, except that the **SEND** and **RECEIVE** commands are issued to the microcomputer and the mainframe respectively. After transfer, a copy of the portable file should now be in your directory on the mainframe computer. Hopefully this process will prove successful even on the first occasion. If there are problems you should consult the *SPSS/PC+ Manual* (Section F) or seek advice from your computer advisory service.

16.3 Expanding your SPSS/PC+ repertoire

This is an introductory book. It is designed to enable first-time users of the SPSS system to become acquainted with the fundamentals of the microcomputer-based SPSS/PC+ package. No attempt has been made to present a comprehensive account of even the base system. Many of the more complex aspects have been presented in a cut-down form or avoided (especially if they are seldom used), and certain procedures have been completely omitted. Also, no attempt has been made to cover the many further procedures available within the various enhancements which have recently become available. As your familiarity with SPSS/PC+ develops you will certainly wish to explore some of these.

The final part of this chapter provides an indication of some aspects of SPSS/PC+ that you might wish to add to your repertoire with the aid of the *SPSS/PC+ Manual* and the written materials provided with the enhancements. The intention is to offer just a brief description so that you can decide for yourself which additional facilities are likely to be of the most interest and benefit to you.

The base system

Procedure commands which have not been discussed in this book are **DESCRIPTIVES, AGGREGATE, REGRESSION** and **WRITE**.

DESCRIPTIVES computes descriptive statistics for variables. Most of the facilities it provides are also available with the **FREQUENCIES** procedure (and the **MEANS** procedure), although it also has a useful **OPTION** for adding standardized scores (i.e. Z scores) to the active file.

AGGREGATE summarizes information about particular groups of cases from the active file, and creates a new active file or system file which contains one summary case per group. Thus with just one break variable, e.g. *sex*, the aggregate file would be likely to consist of just two cases. Summary statistics can be provided for each of these summary cases (for example, **N, SUM, MEAN,** and various percentage functions).

REGRESSION performs multiple regression analyses, calculating the relevant equations and producing a variety of plots. A number of alternative regression methods are available. The procedure can be used with matrix materials (for example, with a results file obtained from the **COR-RELATION** procedure) as the input.

WRITE writes the cases within the active file to an ASCII results file "SPSS.PRC" (this default name may be changed by a **SET** command). The format of the results file can be changed, and the user can specify which variables and which cases should be included in the new file. As the procedure is executed, a screen display of the file format is presented.

Transformation commands which have not been discussed in this book are **COUNT** and **WEIGHT**.

COUNT is a transformation which creates a new variable for each case by counting across several variables. Thus if 20 variables (Q1 to Q20) were coded 1 for 'yes', 2 for 'don't know' and 3 for 'no', a new variable (*dontknow*) could be created to record the number of times the subject ticked 'don't know':

```
SPSS/PC: COUNT dontknow = q1 to q20 (2).
```

WEIGHT is a transformation which affects a system variable *$WEIGHT*. The initial value of this is 1.00 for each case. The command tells the system to assign different weightings to cases. Thus the command **WEIGHT BY SEX** would, if males were coded 1 and females 2, replicate each female case twice in subsequent procedures, thus inflating the size of the sample (the effect of such a change in weighting could be observed by issuing a subsequent **FREQUENCIES** command). The *$WEIGHT* variable may be re-set to 1.00 during a session by issuing the command **WEIGHT OFF**. Otherwise *$WEIGHT* remains as a system variable and will be **SAVED** within a system file or within an **EXPORTED** portable file.

An **operation command** which has not been discussed in this book is **EXECUTE**.

EXECUTE is an operation command which allows the user to run a procedure from an external program (such as **DOS**) **during** an SPSS/PC+ session. Certain problems, however, may prevent this from being an efficient way in which to operate. Accessed while SPSS/PC+ is still active, there is limited memory space available for other programs, and the use of certain programs will cause SPSS/PC+ to abort. Because of such difficulties, all of the examples given in this book of running **DOS** commands to copy or print a file have followed the strategy of first ending the SPSS/PC+ session. A discussion of the use of **EXECUTE**, and of the problems associated with its use, is provided in the *SPSS/PC+ Manual*, pp. C55–C58.

The optional enhancements

The base system, with which this book has been solely concerned, contains software for handling data and files, and includes a number of statistical and report procedures. Since it was introduced (in July 1984) several enhancements have become available. These are sold separately. Your department or computer centre may have none, some or all of these.

Advanced statistics

This includes a number of procedures which provide the means for flexible and powerful multivariate analysis. The types of analysis that can be performed are factorial analysis, loglinear analysis, discriminant analysis, cluster analysis and manova. The last of these considerably extends the range of analysis of variance designs which can be analysed by SPSS/PC+.

In particular it allows the analysis of designs in which the same subject has contributed data to more than one condition (repeated measures).

Tables
This enhancement allows the user to produce high-quality summary tables which can easily be incorporated in final reports, or used as the source for making slides and overhead projection materials. **TABLES** gives the user great freedom in determining the format and content of tables. Titles and footnotes can occupy several lines. The enhancement also allows the examination of questions for which individual cases have provided a number of answers. **TABLES** also allows several tables to be combined within a single display.

Graphics
This enhancement enables the user to produce high-quality graphics to display features of the data. For example, the pie charts procedure will enable the frequencies for various values of a variable to be displayed as suitably sized segments of a circle, each appropriately labelled and shaded with a different design. Pies can be rotated to any angle and the segments can be arranged in any order. A similar degree of flexibility is possible with bar charts, line graphs and scatterplots. With suitable printer facilities, colour graphics can also be produced. The user is able to switch instantly between alternative ways of presenting the data and thus to choose which type of display presents the best portrayal of the relevant features of the dataset.

Data entry
This provides an easy way of entering data and has the special feature that the valid range for any variable can be pre-set. This means that it is easy to identify, at the data entry stage, any datum that is outside this range. This helps in producing a dataset which is clean, i.e. which contains no errors. **DATA ENTRY** also helps in editing data files.

16.4 A final word

The purpose of this book has been to introduce SPSS/PC+ in the style of a tutorial rather than to provide a comprehensive manual. Short cuts have been taken, and some statistically advanced topics such as **REGRESSION** have been avoided. Having covered the material available in this introduction, however, you should be able to follow the detailed presentation of all of the topics covered in the *SPSS/PC+ Manual* and in the manuals provided with each of the various enhancements.

It is likely that in the next few years the SPSS/PC+ system will develop further, and that several new enhancements will become available. The core of the SPSS/PC+ package, however, is unlikely to change, and the skills and concepts you have learned in working through this book should therefore remain useful to you for a long way into the future.

Appendix A:
Installing SPSS/PC+ on a Microcomputer

The SPSS/PC+ system is delivered to the purchaser on a number of diskettes and the system must be installed on the hard disk of a suitable **IBM** (or compatible) microcomputer. This appendix provides instructions for installing and testing the system.

The following instructions for installing SPSS/PC+ assume that:

- the SPSS/PC+ diskettes will be copied from the source (drive A)
- the destination disk will be a hard disk (drive C)
- the computer system boots (i.e. loads the disk operating system, or **DOS**) automatically from the hard disk (drive C)

SPSS/PC+ is sold in diskette form – the number of diskettes will depend on whether only the base system of SPSS/PC+ has been purchased or whether one or more of the optional enhancements (*SPSS/PC+ Advanced Statistics*; *SPSS/PC+ Tables*, etc.) have also been acquired. The following example of an installation assumes that both the base system and *Advanced Statistics* are to be installed.

A **key diskette** is provided. This diskette must be in drive A during certain phases of the operation of SPSS/PC+ and it will need to be authorized during the initial installation process.

- the SPSS/PC+ Base System is stored on diskettes B1 to B9
- SPSS/PC+ *Advanced Statistics* is stored on diskettes A1 to A6

Instructions for loading the SPSS/PC+ diskette information onto the hard disk are clearly given with the disk package and the procedure is directed by instructions displayed on the screen. Considerable care needs to be taken when following these instructions, particularly when the key diskette is being authorized.

The diskettes are inserted, one at a time, into the diskette drive (drive A) and the contents copied onto the hard disk (drive C).

For efficient use, the hard disk is usually divided into many different logical spaces, or directories. The basic root directory is entered when the machine is first switched on. From this there will extend a tree structure with major directories, sub-directories, and possibly even sub-sub-directories. Each major user of the system will generally create a new directory and work from within this. SPSS/PC+ should be installed in a separate directory. If the sequence of instructions provided below is followed, the result will be a directory on the hard disk (named "**\SPSS**") containing all of those parts of the SPSS/PC+ system which have been installed.

Access from any other directory to the SPSS directory can be made using a type of command known as a path command. If SPSS/PC+ is to be easily accessible to all users of the machine (and thus from many different directories) it is useful to include a suitable path command in a file named "AUTOEXEC.BAT". Any instructions included in this file will be executed automatically after the machine has been switched on.

For the efficient operation of SPSS/PC+ the computer system will also need to be specially configured, i.e. customized with instructions about the use of files and buffers (temporary storage locations). This is done by writing two simple commands which can be stored as a batch file called "CONFIG.SYS".

An "AUTOEXEC.BAT" and/or a "CONFIG.SYS" file may already be in operation on your machine. The instructions below assume there are no files with these names already present on the system on which you will be installing SPSS/PC+. If such files *are* in use, you should consult with other users of the machine so that the files can be amended to allow efficient operation of the SPSS/PC+ system without affecting other users' needs.

A test routine for the base system (and similar routines for the optional SPSS/PC+ enhancements) takes the installed system through its paces and, on the initial testing, allows the user to authorize the key diskette. Such authorization is a once only procedure.

Finally, considerable saving can be made on the space occupied by SPSS/PC+ on the hard disk by deleting (perhaps as a temporary measure) some of the rarely used elements of the SPSS/PC+ system. The process of removing and re-installing modules is made easy by the provision of a procedure called 'SPSS MANAGER'.

The process of installing SPSS/PC+ may be seen in terms of a sequence of six stages. In the sections which follow, you will be instructed about how to:

- make diskette copies of the original SPSS/PC+ diskettes
- make a SPSS directory on the hard disk and copy the contents of the SPSS/PC+ diskettes into this directory
- write "AUTOEXEC.BAT"
- write "CONFIG.SYS"
- test the system and authorize the key diskette
- save hard disk space with SPSS MANAGER

1. Making copies of the original diskettes

It is essential to make at least one diskette copy of each of the SPSS/PC+ diskettes (but *not* the key diskette) in addition to the hard disk working copy. Diskettes frequently become damaged or corrupted. Hard disks, too, can fail, and may have to be replaced, or someone may accidentally erase part or all of the contents. The system is initially set up in such a way that the simple command 'format' will effectively wipe clean the hard disk. If steps to prevent this have not already have been taken for your system, you should seek advice (it is a fairly simple matter for an expert to change the **DOS** FORMAT procedure so that this will not happen).

It is always important to keep a backup copy of any file you create for use by SPSS/PC+, but it is of crucial importance to keep at least one copy of each of the original SPSS/PC+ diskettes, except the key diskette *which you should not attempt to copy*. Each backup diskette should be clearly labelled.

Making a copy of a diskette is easy with the **DOS** command **DISKCOPY**.

For machines with two diskette drives

If you are working with a machine with **two** diskette drives they will be drive A and drive B. The (single) hard disk is always drive C. To copy a whole diskette, use the instruction:

```
C>DISKCOPY A: B:     <RETURN>
```

You will be given an instruction to place the source diskette (i.e. the diskette to be copied) in drive A and a new destination (or target) diskette in drive B. Be sure that the destination diskette does not contain valuable information, because everything on it will be overwritten during the copy process. Press any key and the backup copy will be made. You will then be asked whether you wish to copy another diskette. Make any further copies you need.

For machines with one diskette drive

If you are working with a machine with a **single** diskette drive then this drive will act as both drive A **and** drive B. The (single) hard disk is always drive C. To copy a whole diskette, use the instruction:

```
C>DISKCOPY A: A:     <RETURN>
```

Place the source diskette in drive A and press any key, and the machine will copy the contents into memory. The message '**Insert destination disk-ette in Drive A**' will then be displayed. Replace the source diskette with the destination diskette and strike any key. The system will copy each of the files in memory to the destination diskette, thus producing a back-up copy of the source diskette. After the 'Copy Complete' message you will be asked whether you wish to copy another diskette. Make any further copies you need.

2. Making the "SPSS" directory and copying SPSS/PC+ to the hard disk

When you first switch on the machine you will be in the root directory. You need to create another directory called "\SPSS" and to load the information on the SPSS/PC+ diskettes into this directory. The base system diskettes (B1 to B9) should be loaded first. Place diskette B1 in Drive A and type:

```
A:MAKESPSS A:  C:\SPSS    <RETURN>
```

The system will start to copy files from B1 into a (newly created) SPSS directory on the hard disk (drive C).

After all of the files have been copied, you will receive a screen message telling you to insert B2. Do this, and then press any key. Files will be copied from B2 to the hard disk.

Continue in this way for all of the disks labelled B3 to B9.

At certain points in this process you will be asked whether certain modules should be copied, and advised on the amount of space which these modules will take on the hard disk. The choice is yours; if there are unlikely to be problems of disk space it is best to copy all of the modules initially. Enter Y for yes, and then <RETURN> in response to the prompt. The procedure described here assumes that all of the modules will initially be loaded onto the hard disk. Directions for removing unwanted modules will be given later.

After copying the diskette B9 you will be asked whether you wish to install *Advanced Statistics*. If you do, then you should load diskettes A1 to A6.

3. Writing "AUTOEXEC.BAT"

A suitable path instruction will allow access to the SPSS/PC+ system (and to **DOS**) from any directory. For the greatest ease of operation such an instruction can be included in a command (written into a 'batch file') which is automatically executed when the machine is switched on. Only one such batch file is permitted. It must be called "AUTOEXEC.BAT", and it must reside in the root directory.

The instructions given below assume there to be no existing

"AUTOEXEC.BAT" file. If other users have already set a file with this name to perform other functions, they should be consulted before you attempt to create your own "AUTOEXEC.BAT". Only one such file can be resident on the machine at any one time, and if you create a file with this name it will overwrite any existing "AUTOEXEC.BAT".

To write an "AUTOEXEC.BAT" file, follow these instructions:

(a) Starting from **DOS** within the root directory, and in response to the 'C>' prompt, type the instruction:

```
C> type autoexec.bat      <RETURN>
```

(b) If you are presented with a screen display showing one or more command lines, you know that an "AUTOEXEC.BAT" file already exists. If this is the case then consult with others before proceeeding.

If, however, the message 'File not present' appears you will know that there is no existing "AUTOEXEC.BAT" file.

(c) If there is no existing file, enter the following commands:

```
C> copy con autoexec.bat        <RETURN>
C> path \;\spss;\dos            <RETURN>
```

(this assumes that **DOS** is located in a directory named 'DOS')

Now issue a <Control Z> instruction and then press <RETURN> once more. The "AUTOEXEC.BAT" file will now be written to disk.

The path command contained within "AUTOEXEC.BAT" will not be in operation until either the reset button is pressed, or the machine is switched on again (do not do this yet). Thereafter, the path command will be issued automatically each time the machine is switched on. Any user will then be able to access the SPSS directory, and the SPSS/PC+ system, from any directory simply by typing 'spsspc' in response to the C> prompt. The various **DOS** commands will also be accessible from any directory.

As you become more knowledgeable and more proficient in using the operating system you may wish to revise "AUTOEXEC.BAT" to include other useful automatic commands. But to ensure that the SPSS/PC+ system can be easily accessed from any directory, any subsequent revision should retain an instruction equivalent to the above path command.

4. Writing "CONFIG.SYS"

To ensure efficient operation of SPSS/PC+ the computer system will need to be configured (or 'customized') appropriately. The number of file buffers needs to be increased to 8, and the number of files which can be open simultanously needs to be increased to 20. Commands specifying these changes should be included in a **DOS** file called "CONFIG.SYS". Like "AUTOEXEC.BAT", this file will be executed automatically when the machine is switched on.

The instructions given below will produce a suitable "CONFIG.SYS" file after testing to see whether such a file already exists. If other users have written such a file for other purposes, they should be consulted before you attempt to create your own file of that name. Only one "CONFIG.SYS" can be resident on the machine at any one time, and if you create such a file it will overwrite any existing "CONFIG.SYS".

If, however, the machine is to be used principally as a SPSS/PC+ machine, the following sequence will create the appropriate file:

(a) Starting from **DOS** within the root directory, and in response to the 'C>' prompt, type the instruction:

```
C> type config.sys        <RETURN>
```

(b) If you are presented with a screen display listing some commands, you know that a "CONFIG.SYS" file already exists. If this is the case then consult with others before proceeding. If, however, the message '**File not present**' appears you will know that there is no existing "CONFIG.SYS" file.

(c) If there is no existing file, enter the following commands:

```
C> copy con config.sys           <RETURN>
C> files=20                      <RETURN>
C> buffers=8                     <RETURN>
```

Now issue a <Control Z> instruction and then press <RETURN> once more. The "CONFIG.SYS" file will now be written to disk.

The parameter settings specified within the "CONFIG.SYS" file will not be in operation until you press the reset button or switch on the machine on a future occasion. Thereafter, the two commands within this file will automatically configure the system appropriately.

5. Testing the system

When "AUTOEXEC.BAT" and "CONFIG.SYS" have been written, and the machine re-booted (i.e. the <RESET> pressed or the machine switched on again) you are ready to test the system and to authorize the key diskette. The write-protect tab on the side of the key diskette should be **temporarily** removed at this stage.

A test job is included within the base system (and appropriate test jobs are also included within the *Advanced Statistics* and other enhancements).

To create your own directory and to use the test procedures:

(a) Switch on the machine; you will be in the root directory and the **DOS** flag 'C>' will be displayed

(b) Make a user directory, including a short version of your name (one word), with a command e.g. 'md\johndoe', e.g.

```
C> md\johndoe    <RETURN>
```

(c) When you enter the root directory, on future occasions, you will have to change to your own directory using the 'cd\' command. Do this now, e.g.

```
C> cd\johndoe    <RETURN>
```

(d) To run the base SPSS/PC+ test job **from your own directory**, enter

```
C>spsspc spss\basetest.inc    <RETURN>
```

The SPSS/PC+ logo will be displayed and the system will run through a series of trials of the modules included within the base system. The message MORE will frequently appear in the top right-hand corner. In response to this, you should press any key. At the end of the tests you will be returned to **DOS** (and to your own directory).

This assumes that all of the modules have in fact been installed. If they have not then you will need to change "BASETEST.INC" (consult the *SPSS/PC+ Manual*).

If you have installed *Advanced Statistics* and/or other enhancements you will need to test these, too. This is especially important because the test routines include the opportunity to authorize your key diskette to unlock these systems. The optional systems cannot be used unless the key diskette has been authorized in this way, and such authorization is a once only procedure. Run the tests using the same procedure as (4) above, but instead of "BASETEST.INC" use "ADVTEST.INC", etc.

Read the screen information carefully and follow the instructions for authorizing the key diskette. When authorization is complete **replace the write-protect tab** on the key diskette.

The system should now be fully installed and working correctly. When you wish to use SPSS/PC+ you simply:

(a) Switch machine on (and wait for system checks)

(b) Respond to the **DOS** flag ('C>') by changing from the root directory to your own directory:

```
C>cd\johndoe    <RETURN>
```

(c) Check that the key diskette is in drive A

(d) Enter the command 'spsspc' to enter the SPSS/PC+ system

```
C>spsspc        <RETURN>
```

The SPSS/PC+ logo will appear and you will be presented with the 'SPSS/PC:' prompt.

The SPSS/PC+ session has now begun and the system is awaiting your SPSS/PC+ command.

6. Saving space with 'SPSS MANAGER'

The full SPSS/PC+ package takes a great deal of disk storage space, especially if several of the enhancements are added. Certain of the procedures may be rarely used, however, and provision has therefore been made for the easy removal and re-installation of some modules from the hard disk. This is achieved with the procedure SPSS MANAGER.

The hard disk copy of a procedure is a further back-up to any diskette back-up that has been made. Removing a procedure from the hard disk therefore removes one safety layer; to compensate for this an *additional* diskette copy of a procedure should be made when the procedure is removed from the hard disk.

The command:

```
SPSS/PC: SPSS MANAGER STATUS.
```

produces a display listing the non-permanent SPSS/PC+ procedures currently installed on your system. Information is also provided about the space occupied by these procedures and the unused disk space currently available.

To remove a module

The unit for removal or installation is the module, and the base system includes four non-permanent modules, each containing one, two or four procedures. You will have to retain a whole module if it contains at least one procedure to which continued access is necessary.

For example, the command:

```
SPSS/PC: SPSS MANAGER REMOVE=AGGREGATE;ANOVA.
```

will produce a display of all of the eight procedures included in the two modules containing the procedures AGGREGATE and ANOVA, and will indicate how much extra disk space will become free if both of these modules are removed. You will be asked whether you wish to proceed with the removal operation. If you confirm your request (by entering Y and <RETURN>) the system will remove the specified module(s) and issue a '**Removal complete**' message.

To install a module

The command:

```
SPSS/PC: SPSS MANAGER INSTALL=ONEWAY.
```

will list the two procedures (ONEWAY and REGRESSION) included in the relevant module. The display also includes a note of the source from which the module(s) to be installed will be taken (by default this is the diskette drive A) and the destination (usually 'c:\spss\', i.e. the "\spss" directory on the hard disk, drive C).

If the source is another drive (for example, drive B) then it must be identified using a From subcommand. For example:

```
SPSS/PC: SPSS MANAGER INSTALL=ONEWAY /FROM "B:".
```

If the source is a second hard disk (drive D), rather than a diskette drive, the directory containing the module should be identified with a path command; e.g.:

```
SPSS/PC: SPSS MANAGER INSTALL=ONEWAY /FROM "D:\SPSSSUP".
```

As for the REMOVE procedure, the INSTALL command will produce an initial display of the procedures included in the modules containing any procedures specified for installation.

The display will also report on the space taken up by each procedure and the space currently available on the hard disk. If you confirm the initial request (by replying to the '**Do you wish to proceed?**' message with 'Y' and <RETURN>) you will be informed of the diskette which needs to be placed in the drive. Pressing <RETURN> again will cause the relevant information to be copied from the diskette to the hard disk.

Modules can be removed and reinstalled as often as you wish.

Appendix B: Using Editors Other Than REVIEW to Create SPSS/PC+ Files

There are many advantages of using the SPSS/PC+ editor, **REVIEW**, for editing SPSS/PC+ files, particularly if you need to edit a file within a session. The process of writing long data files or data definition files, however, can take a considerable time, and if there is pressure on the SPSS/PC+ facility it might be necessary to write such files using another editor. Many word-processing systems are suitable, and spreadsheet and database management systems can also be used.

Files for use by SPSS/PC+ should be kept as plain as possible, avoiding any special control facilities which the package might offer. The files should consist of alphanumeric text set in lines of no more than 80 columns, with <RETURN> used to end each line and <Control Z> to end the file. For a spreadsheet or database system to be useful you will need to be able to obtain an alphanumeric version of the data content in a plain form, without headings, special format instructions and the like. The manual for the particular program may give you instructions on how to produce such files.

If you are familiar with a particular editor you might like to try composing a simple data definition file (including a number of lines of dummy data) to see whether the result proves acceptable to SPSS/PC+. If you encounter difficulties, try editing the file with **REVIEW**. Sometimes the removal of obvious blemishes within the file will be sufficient to make the contents usable by the system. If difficulties persist, your computer advisor may be able to help you.

You may be able to write files on a computer which is not itself **IBM/PC** compatible. Several commercially available programs enable text files written in one disk format to be transferred to IBM/PC format. Thus you could write the file using your preferred editor on a readily available non-compatible machine and then transfer the completed file to an IBM/PC format disk. This disk would then be inserted into the diskette drive A of the **IBM** machine and the appropriate file(s) copied into the appropriate directory of drive C (the hard disk). The SPSS/PC+ system would then access the file(s) and process the information. Such files can also be re-edited with **REVIEW**.

It is not possible to include here details of the many editors which can be used to produce files for SPSS/PC+. The use of one such editor, however, will be briefly described. WordStar is a widely available word-processing package, and is used on both **IBM** compatible and other machines.

Assuming that you are familiar with the normal word-processing use of WordStar, how should this editor be used to produce SPSS/PC+ files? There are two rules: use the non-document file mode and keep to text material (avoid 'controls').

(a) When preparing a text file with WordStar it is usual to take advantage of the document file mode. To edit a new or existing file a D command is issued. The document mode sets a line length (providing an automatic word-wrap at the end of a line) and makes page breaks as appropriate.

Such facilities, however, are unnecessary when creating a file for use by SPSS/PC+ and will produce errors. Thus the alternative non-document mode is used. To create or edit a non-document file you enter the command N (rather than D).

You may now type normally and use the normal editing controls, but there will be no automatic <RETURN> at the end of a line, and no automatic page break. Keep all lines to a maximum of 80 columns and issue a <RETURN> command at the end of each line.

(b) When using WordStar to produce documents, several print commands may be used. Some are special instructions to the printer (dot commands) and others are inserted within the text material to control such features as underlining and bold printing. Do not use commands of these types when preparing a file for use by SPSS/PC+. Such commands will serve no useful purpose and will cause processing errors.

The completed file – a matrix of numbers constituting a data file perhaps, or a set of statements constituting a data definition file – can now be saved using the usual command <^KD>.

The saved non-document file can now be transferred to the directory to be used for SPSS/PC+ analysis. If the disk format used was not of IBM/PC standard then the file will first need to be copied onto an appropriately formatted diskette. Programs such as Media-Master are available for this purpose.

When the IBM/PC compatible version of the file has been produced, the diskette is inserted into **drive A** of the SPSS/PC+ machine. It should then be copied into the appropriate directory of the hard disk (**drive C**). The file (call it "**wsgend.def**") can be copied from **drive A** to the current directory of **drive C**, by issuing the **DOS** command:

```
A> copy a:wsgend.def c:        <RETURN>
```

This will save a copy of the file (also named "**wsgend.def**") on the hard disk. The SPSS/PC+ key diskette can now be inserted into **drive A**, and SPSS/PC+ entered (by typing 'C:spsspc'). When the SPSS/PC: prompt is presented, the instruction 'include "wsgend.def" ' will cause the file originally created in WordStar to be read by SPSS/PC+. Remember that if the data are contained in a separate data file this must also be copied into the directory so the system can read the data defined in the definition file.

If problems are encountered, the file can be re-edited using **REVIEW**. If, in error, you have used the document mode to prepare a file for SPSS/PC+, it may be possible to make the file acceptable to SPSS/PC+ by editing out the page break symbol and any other odd symbols.

Appendix C:
Gender Attitudes Survey Questionnaire

The questionnaire for the gender attitudes survey was designed specifically to collect data that would allow many features of the SPSS/PC+ system to be illustrated. The 50 respondents were all psychology students from one British university. All students attending psychology classes on two particular days were asked to complete the questionnaire once. The 176 completed questionnaires were then numbered and a table of random numbers used to select 50 cases for inclusion in the sample. This limitation was made to reduce the data set to a size which readers of this book could enter into a file without undue effort (it takes about three-quarters of an hour to create a file containing the complete set of data).

The questionnaire begins by asking for certain personal details (sex, age, faculty and year of study) and then asks the respondent to indicate the degree of agreement or disagreement with each of 10 statements. These statements were derived from preliminary discussions about gender issues with a small group of students.

There are then 10 first-person statements about emotions and interpersonal relationships, and the respondent is asked to judge whether more men or women would be likely to agree with each of these statements. Responses to these items may be interpreted as reflecting stereotypes about men and women. Later in the questionnaire (and over the page, to reduce the likelihood that respondents might refer back to their former responses) the same set of statements is presented. This time, however, the subject is asked whether the statement does or does not apply personally. By examining the proportions of males and females agreeing with each of these items, some indication is gained of whether any stereotypes which emerge actually reflect the personal responses of the young men and women who took part in the survey.

Following a question about marital status, unmarried people are directed to a set of items designed to assess the importance attached to various characteristics for someone who would be a desirable date. Again, these items are repeated over the page, and on the second occasion the respondent is asked to indicate the importance attached to the characteristics for someone who might be a desirable marriage partner. A comparison between the two sets of responses might indicate that the importance of particular features differs somewhat for a desirable temporary and permanent partner.

Questions on weight, height and expected salary were included in order to provide data that (like age) could be regarded, non-controversially, as true numbers rather than as codes, thus permitting analyses which demand that data be truly numerical.

The 12 questions from the short form of the Eysenck Personality Inventory (Eysenck, 1958) were included to enable two personality characteristics – extraversion and neuroticism – to be measured. These questions are not included in the questionnaire reproduced in this appendix, but the relevant data are included in Appendix E.

SEX (circle one): M / F   AGE: .... (yrs)

Indicate (by circling the appropriate number) your **Faculty** of study:

1 – ARTS   2 – SCIENCE   3 – SOCIAL STUDIES

Year of study (circle one): First   Second   Third

Please circle the number beside each of the statements below to
indicate how much you agree or disagree:

*Please
leave
blank*

ID _ _ _ _

SEX _

AGE _ _

FAC _

YR _

	Strongly Agree				*Strongly Disagree*	
1. Women are equal to men in every way	1 2 3 4 5 6 7					VAR001 _
2. New legislation is needed to ensure equal rights	1 2 3 4 5 6 7					VAR002 _
3. Many women use 'sex discrimination' as an excuse for their own shortcomings	1 2 3 4 5 6 7					VAR003 _
4. There are many activities in which men are clearly superior to women	1 2 3 4 5 6 7					VAR004 _
5. The women's movement has had little serious effect	1 2 3 4 5 6 7					VAR005 _
6. The legal system often treats men more harshly than women	1 2 3 4 5 6 7					VAR006 _
7. Women are emotionally more sensitive than men	1 2 3 4 5 6 7					VAR007 _
8. Generally speaking, men have a higher level of sexual desire than women	1 2 3 4 5 6 7					VAR008 _
9. Most men make a much less serious commitment to marriage than women	1 2 3 4 5 6 7					VAR009 _
10. Most men feel very threatened by highly intelligent women	1 2 3 4 5 6 7					VAR010 _

| If a survey were conducted with a very large 'representative' sample of men and women (in equal numbers), do you think that more **men** or more **women** would **agree** with each of the following statements (circle MEN or WOMEN against each item): | *Please leave blank* |

(circle one)

I often feel a little depressed	MEN	WOMEN	ST01 __
I seem to think about sex a lot of the time	MEN	WOMEN	ST02 __
I often feel a little anxious	MEN	WOMEN	ST03 __
I have to admit that I get angry quite often	MEN	WOMEN	ST04 __
I enjoy figuring out abstract problems	MEN	WOMEN	ST05 __
I enjoy taking risks	MEN	WOMEN	ST06 __
I frequently have trouble sleeping	MEN	WOMEN	ST07 __
I tend to try to dominate other people	MEN	WOMEN	ST08 __
I am often bored	MEN	WOMEN	ST09 __
I find it difficult to trust people	MEN	WOMEN	ST10 __

What is your current marital status? (tick one)

    Single   /   Married   /   Divorced   /   Separated   /   Widowed   MARIT __

    ...   .

If you are **MARRIED** do **NOT** answer the following question but turn to the next page.

For unmarried people only

What do you look for in a desirable date? Using the list of attributes printed below, indicate the importance you would attach to each when thinking about dating someone.

V=Very important	I=Important
N=Not important	U=Undesirable

It is (V or I or N or U) that my date should have:

Ambition	V	I	N	U	DAT01 __
Intelligence	V	I	N	U	DAT02 __
Physical attractiveness	V	I	N	U	DAT03 __
A sense of humour	V	I	N	U	DAT04 __
Kindness	V	I	N	U	DAT05 __
Sexual experience	V	I	N	U	DAT06 __
Good morals	V	I	N	U	DAT07 __
Dependability	V	I	N	U	DAT08 __

Statements which you have considered previously are presented once more below. This time you should consider whether each of the statements applies to **you** yourself. Circle either TRUE or FALSE to indicate whether the statement is true or false for **you**.	*Please leave blank*

(circle one)

I often feel a little depressed	TRUE	FALSE	PRS01 __
I seem to think about sex a lot of the time	TRUE	FALSE	PRS02 __
I often feel a little anxious	TRUE	FALSE	PRS03 __
I have to admit that I get angry quite often	TRUE	FALSE	PRS04 __
I enjoy figuring out abstract problems	TRUE	FALSE	PRS05 __
I enjoy taking risks	TRUE	FALSE	PRS06 __
I frequently have trouble sleeping	TRUE	FALSE	PRS07 __
I tend to try to dominate other people	TRUE	FALSE	PRS08 __
I am often bored	TRUE	FALSE	PRS09 __
I find it difficult to trust people	TRUE	FALSE	PRS10 __

What would you look for in a desirable marriage partner? Consider each of the attributes printed below, and indicate the importance you would attach to each when thinking about marrying someone.

V=Very important   I=Important
N=Not important   U=Undesirable

It is (V or I or N or U) that someone I would MARRY should have:

Ambition	V	I	N	U	MAR01 __
Intelligence	V	I	N	U	MAR02 __
Physical attractiveness	V	I	N	U	MAR03 __
A sense of humour	V	I	N	U	MAR04 __
Kindness	V	I	N	U	MAR05 __
Sexual experience	V	I	N	U	MAR06 __
Good morals	V	I	N	U	MAR07 __
Dependability	V	I	N	U	MAR08 __

Now please answer the following questions:

HTF __
HTI __ __

What is your HEIGHT?   ....ft ....inches   WST __ __

What is your WEIGHT?   ....st ....lb   WLB __ __

If you were to get a job immediately after completing your current degree, what salary would you **expect** to get in your **first** year (i.e. over the first 12 months). Do NOT allow for inflation but give the salary at present money values: £..............

SALARY

__ __ __ __ __

E1 __	E2 __
E3 __	E4 __
E5 __	E6 __
E7 __	E8 __
E9 __	E10 __
E11 __	E12 __

NOW PLEASE COMPLETE THE LAST SET OF QUESTIONS, OVERLEAF

Appendix D:
Gender Attitudes Survey Codebook

The next pages contain a full codebook for the gender attitudes survey, including the column number occupied by the variable (1–80), the type of data (N = numerical; C = categorical), the valid values and 'missing' value, and any value and variable labels.

Several additional variables (*neur, extra, sexism, sextract* and *problem*) were created from the original data set. Appendix F includes the **COMPUTE** commands used to create these variables. For some of the analyses included in this book, the values for the original data and for new variables were recoded using an appropriate **RECODE** command. See the text and Appendix F for further details.

Variable name	Col/s	Type	Values	Value labels	Miss.	Var. label
ID	1–4	N	–	–	9999	Subject no.
SEX	5	C	1	Male	9	Subject sex
			2	Female		
AGE	6–7	N	–		99	Subject age
FAC	8	C	1	Arts	9	Home faculty
			2	Science		
			3	Soc. studies		
YR	9	C	1	1st year	9	Year of study
			2	2nd year		
			3	3rd year		
VAR001	10	?	1 to 7	Strongly agree/ Strongly disagree	9	Women equal to men
VAR002	11	?	1 to 7	Strongly agree/ Strongly disagree	9	New legislation needed

Variable name	Col/s	Type	Values	Value labels	Miss.	Var. label
VAR003	12	?	1 to 7	Strongly agree/ Strongly disagree	9	Women use sd as excuse
VAR004	13	?	1 to 7	Strongly agree Strongly disagree		Men superior in many activities
VAR005	14	?	1 to 7	Strongly agree/ Strongly disagree	9	Women's move. ineffective
VAR006	15	?	1 to 7	Strongly agree/ Strongly disagree	9	Legal system anti-men
VAR007	16	?	1 to 7	Strongly agree/ Strongly disagree	9	Women more sensitive
VAR008	17	?	1 to 7	Strongly agree Strongly disagree	9	Men more sexy
VAR009	18	?	1 to 7	Strongly agree/ Strongly disagree	9	Men less marriage commitment
VAR010	19	?	1 to 7	Strongly agree Strongly disagree	9	Men threat. by intell. women
ST01	20	C	1 2	MEN WOMEN	9	Depressed often (M–W)
ST02	21	C	1 2	MEN WOMEN	9	Sex thoughts often (M–W)
ST03	22	C	1 2	MEN WOMEN	9	Anxious often (M–W)
ST04	23	C	1 2	MEN WOMEN	9	Angry often (M–W)

Variable name	Col/s	Type	Values	Value labels	Miss.	Var. label
ST05	24	C	1 2	MEN WOMEN	9	Enjoy abstract problems (M–W)
ST06	25	C	1 2	MEN WOMEN	9	Enjoy risks (M–W)
ST07	26	C	1 2	MEN WOMEN	9	Trouble sleeping (M–W)
ST08	27	C	1 2	MEN WOMEN	9	Domin. people (M–W)
ST09	28	C	1 2	MEN WOMEN	9	Bored often (M–W)
ST10	29	C	1 2	MEN WOMEN	9	Trust difficult (M–W)
MARIT	30	C	1 2 3 4 5	Single Married Divorced Separated Widowed	9	Marital status
DAT01	31	C	1 2 3 4	Very imp. Important Not imp. Undesirable	9	Ambition (date)
DAT02	32	C		As DAT01	9	Intelligence (date)
DAT03	33	C		As DAT01	9	Phys. attract. (date)
DAT04	34	C		As DAT01	9	Sense of hum. (date)
DAT05	35	C		As DAT01	9	Kindness (date)
DAT06	36	C		As DAT01	9	Sexual exper. (date)
DAT07	37	C		As DAT01	9	Good morals (date)
DAT08	38	C		As DAT01	9	Dependability (date)
PRS01	39	C	1 2	TRUE FALSE	9	Depressed often (SELF)
PRS02	40	C	1 2	TRUE FALSE	9	Sex thoughts often (SELF)
PRS03	41	C	1 2	TRUE FALSE	9	Anxious often (SELF)
PRS04	42	C	1 2	TRUE FALSE	9	Angry often (SELF)

Variable name	Col/s	Type	Values	Value labels	Miss.	Var. label
PRS05	43	C	1 2	TRUE FALSE	9	Enjoy abstract problems (SELF)
PRS06	44	C	1 2	TRUE FALSE	9	Enjoy risks (SELF)
PRS07	45	C	1 2	TRUE FALSE	9	Trouble sleeping (SELF)
PRS08	46	C	1 2	TRUE FALSE	9	Domin. people (SELF)
PRS09	47	C	1 2	TRUE FALSE	9	Bored often (SELF)
PRS10	48	C	1 2	TRUE FALSE	9	Trust difficult (SELF)
MAR01	49	C	1 2 3 4	Very imp. Important Not imp. Undesirable	9	Ambition (spouse)
MAR02	50	C		As MAR01	9	Intelligence (spouse)
MAR03	51	C		As MAR01	9	Phys. attract. (spouse)
MAR04	52	C		As MAR01	9	Sense of hum. (spouse)
MAR05	53	C		As MAR01	9	Kindness (spouse)
MAR06	54	C		As MAR01	9	Sexual exper. (spouse)
MAR07	55	C		As MAR01	9	Good morals (spouse)
MAR08	56	C		As MAR01	9	Dependability (spouse)
HTF	57	N		—	9	Height (feet)
HTI	58–59	N		—	99	Height (inches)
WST	60–61	N		—	99	Weight (stones)
WLB	62–63	N		—	99	Weight (pounds)
SALARY	64–68	N		—	99999	Annual salary (£ sterling)
E1	69	C	1 2 3	YES NOT SURE NO	9	H/D no reason
E2	70	C		As E1	9	Prefer action

Variable name	Col/s	Type	Values	Value labels	Miss.	Var. label
E3	71	C		As E1	9	Mood w/o cause
E4	72	C		As E1	9	Rapid action
E5	73	C		As E1	9	Moody
E6	74	C		As E1	9	Mind wander
E7	75	C		As E1	9	Make friends
E8	76	C		As E1	9	Quick-sure acts
E9	77	C		As E1	9	Lost in thought
E10	78	C		As E1	9	Lively
E11	79	C		As E1	9	Bubbly/sluggish
E12	80	C		As E1	9	Social contacts

Appendix E: Gender Attitudes Survey Data

The two pages of this appendix contain *all* of the data from the gender attitudes survey which has been subjected to analysis throughout this book. There are **50 cases**. The **69 variables** extend over 80 columns.

The guide-lines printed after each segment of 10 columns are to help with data entry – they should not of course be entered with the data. Column numbers are also given above the data matrix, again to aid data entry (as you enter data, using **REVIEW**, a number at the bottom of the screen indicates the current cursor column).

(Gender Attitudes Survey Data - Cases 0001 to 0020)

```
           1 1111111112 2222222223 3333333334 4444444445 5555555556 6666666667 7777777778
  1234567890 1234567890 1234567890 1234567890 1234567890 1234567890 1234567890 1234567890

0001120122 1636477732 1221122221 3121333322 2211121231 2123226011 1000600011 3313112133
0002220121 1347663622 1211121211 2221133122 2221222222 2113315091 0140520011 3231113211
0003120225 3435524542 1211121221 3211132311 1212121131 1113235111 0021700011 1111231213
0004219122 1476437312 2211111211 2222132111 1222222222 2213215050 8050750013 1231332111
0005226122 1446626312 1212112121 2221132111 2112212111 2112115051 1000650013 2311131121
0006220121 6755465542 9219921991 2221232112 1111121122 2113215020 8000800013 1113112121
0007120135 5635227762 1211122111 1111144121 1212222111 1113116001 1080700013 1113213323
0008120122 2235252652 1221121121 9999999921 1211222221 1113225111 0000650031 3133233313
0009219122 1524527452 2211111221 3231132211 1222221121 3112219990 8129999911 1312131113
0010224122 1612517652 2211132111 2211132111 1222221122 1113215091 0000900013 2131211111
0011227323 2727434432 1212111211 3222232322 2222222232 2223235020 8070700031 3131111121
0012122235 3336325332 1211111211 2211233322 2211112211 2223325101 4000750011 1111121111
0013219113 1476364642 1122212111 1121131112 1121211211 2113115000 8000400031 3131113111
0014245111 2456363562 1299999921 2121132222 1112221221 2113225061 1000500011 1323113113
0015122235 2436333322 1211121211 2121123121 2212212122 2112215101 3020900033 3312131131
0016231132 1345766631 9999999992 9999999922 2111222232 3223225021 0000650032 3131132313
0017220135 1666747762 2291199921 2121133121 1222222221 2231225040 9000650013 1231131111
0018222133 2536434319 9999999991 2111121112 1211222221 1112115010 9080970032 3131113121
0019221117 1655436532 1221121111 2221131112 1211222232 3113115080 8120750031 1331121331
0020124111 1773466772 1221122221 2121233211 1211121221 2123325101 1120700013 1111311313
```

(Gender Attitudes Survey Data - continued - Cases 0021 to 0050)

```
           1 1111111112 2222222223 3333333334 4444444445 5555555556 6666666667 7777777778
 1234567890 1234567890 1234567890 1234567890 1234567890 1234567890 1234567890 1234567890

0021221329 1666426412 2221111211 3221132211 2222221132 2113225091 2006000011 1331332331
0022219322 1777777732 1221121211 3231131111 1122112222 2113115081 1060750011 1113111121
0023229321 1437957752 1222212212 9999999922 1111212211 2113115020 8060700031 3131113231
0024118212 2342213622 1211111121 3222132122 1111291122 2113215091 1020700091 1113111123
0025221312 1445657422 1212121111 9999999911 1221121131 1113325060 9070700021 3131333111
0026148315 7636366432 1212122222 3111132121 2222222231 2241115081 1081000042 2111313333
0027121211 2563563732 1211121111 9999999912 1111211122 2070600032 2070600032 1331131112
0028231334 3336535532 1222222111 3321132221 2222222233 3222225050 8080750011 1111131211
0029120116 1323523342 1222221111 1321132212 1122111112 2232226001 2080400011 3331311111
0030223331 1273177312 1211121211 3222231122 1211111112 1112219990 9021000011 1111211111
0031119335 1412461412 1221122221 2121133122 1211221232 1120055091 1000550011 1311131211
0032221335 1775256632 1221122221 2222232212 2111122122 2223315090 9000700013 3131113123
0033233311 1366257422 1122222221 3211123221 1111122122 2223225060 9020700013 1111131111
0034120313 5222342242 1211121121 2221132121 2111121132 2113315101 1070750011 3111131111
0035121126 4126572412 1212121121 2231123321 1211121132 2113215111 2030850011 3111311311
0036221311 1775357732 1221112121 2231132211 1111111122 2113215111 9030700023 2131113131
0037222235 2325662622 1221112122 9999999922 1111111122 2232225040 8020750011 3133313131
0038132324 2549323432 1211121212 3222132112 2211122222 3113215050 9020800021 2131131112
0039237335 4245344652 2111221212 3211132211 2221121123 2213125061 0050600013 2131113131
0040127325 7331532312 1211121221 2131133211 2122122232 3113215041 0100800011 1113121131
0041221232 1645737632 1121121221 2222132212 1112212212 1213215101 9070720013 1131121311
0042121234 3335335552 1121111221 3222232212 2222222222 1113215040 4000700011 2131112113
0043221225 5436443642 1211111221 2221132211 3122221222 2223116011 9010800011 3121113131
0044120222 1725646662 1211111111 2221222321 2212222222 3123325071 0071200031 1323331111
0045119222 1532723442 1211121211 3111132111 2221122232 2123225062 3070700011 1311312111
0046221213 3545353642 1211121121 3211132111 1221222222 2223226001 8030750011 1311131132
0047122332 1655466442 1299921121 3211132131 1211211222 2123115040 9100700011 1131112111
0048220315 1623612542 1291121221 3211131111 2111222131 2122215070 1000500011 1113111211
0049120312 2655434632 9919999991 3211142212 1212121231 1113115061 2010700011 1131113111
0050126332 1552462562 1111121221 2232142212 1212222122 3223225091 2100800013 3231131213
```

Appendix F: Gender Attitudes Survey Data Definition File

Gender.def

```
DATA LIST FILE= "gender.dat" / id 1-4 sex 5 age 6-7 fac 8 yr 9
var001 to var010 10-19 st01 to st10 20-29 marit 30
dat01 to dat08 31-38 prs01 to prs10 39-48 mar01 to mar08 49-56
htf 57 hti 58-59 wst 60-61 wlb 62-63 salary 64-68
e1 to e12 69-80.
VARIABLE LABELS id "Subject No." / sex "Subject sex"
/ age "Subject age" / fac "Home faculty"
/ yr "Year of study" / var001 "Women equal to men"
/ var002 "New legislation needed"
/ var003 "Women use sexual dis. as excuse"
/ var004 "Men superior in many activities"
/ var005 "Women's movement ineffective"
/ var006 "Legal system anti-men"
/ var007 "Women more sensitive"
/ var008 "Men more sexy"
/ var009 "Men less marriage commitment"
/ var010 "Men threatened by intell. women"
/ st01   "Depressed often (M-W)"
/ st02   "Sex thoughts often (M-W)"
/ st03   "Anxious often (M-W)"
/ st04   "Angry often (M-W)"
/ st05   "Enjoy abstract problems (M-W)"
/ st06   "Enjoy risks (M-W)"
/ st07   "Trouble sleeping (M-W)"
/ st08   "Dominate people (M-W)"
/ st09   "Bored often (M-W)"
/ st10   "Trust difficult (M-W)"
/ marit  "Marital status"
/ dat01  "Ambition (date)" / dat02   "Intelligence (date)"
/ dat03  "Phys attract. (date)" / dat04 "Sense of hum. (date)"
/ dat05  "Kindness (date)" / dat06   "Sexual exper. (date)"
/ dat07  "Good morals (date)" / dat08   "Dependability (date)"
/ prs01  "Depressed often (SELF)"
/ prs02  "Sex thoughts often (SELF)"
/ prs03  "Anxious often (SELF)"
/ prs04  "Angry often (SELF)"
/ prs05  "Enjoy abstract problems (SELF)"
/ prs06  "Enjoy risks (SELF)"
/ prs07  "Trouble sleeping (SELF)"
/ prs08  "Dominate people (SELF)"
/ prs09  "Bored often (SELF)"
/ prs10  "Trust difficult (SELF)"
/ mar01  "Ambition (spouse)" / mar02   "Intelligence (spouse)"
```

```
/ mar03   "Phys attract. (spouse)"/ mar04 "Sense of hum.(spouse)"
/ mar05   "Kindness (spouse)" / mar06 "Sexual exper. (spouse)"
/ mar07   "Good morals (spouse)"/ mar08 "Dependability (spouse)"
/ htf     "Height (feet)"    / hti  "Height (inches)"
/ wst     "Weight (stone)"   / wlb  "Weight (pounds)"
/ salary  "Annual salary ## sterling"
/ E1      "H/D no reason"    / E2   "Prefer action"
/ E3      "Mood w/o cause"   / E4   "Rapid action"
/ E5      "Moody"            / E6   "Mind wander"
/ E7      "Make friends"     / E8   "Quick-sure acts"
/ E9      "Lost in thought"  / E10  "Lively"
/ E11     "Bubbly/sluggish"  / E12  "Social contacts".
VALUE LABELS sex 1 "Male" 2 "Female"
/ fac 1 "arts" 2 "science" 3 "soc. studies"
/ yr  1 "first year"  2 "second year" 3 "third year"
/ var001 to var010 1 "Strongly Agree" 7 "Strongly disagree"
/ st01 to st10 1 "MEN" 2 "WOMEN"
/ marit 1 "Single" 2 "Married" 3 "Divorced" 4 "Separated"
5 "Widowed"
/ dat01 to dat08 mar01 to mar08 1 "Very Imp." 2 "Important"
3 "Not Imp." 4 "Undesirable"
/ prs01 to prs10 1 "TRUE" 2 "FALSE"
/E1 to E12 1 "YES" 2 "NOT SURE" 3 "NO".
MISSING VALUES sex fac yr var001 to htf E1 to E12 (9)
/ age hti wst wlb (99) / id (9999) / salary (99999).
```

F.1 Transformed variables

The following data transformation commands were used to **COMPUTE** new variables and to **RECODE** variables from the gender attitudes survey. See Chapter 8 for general information on transformations. It will probably be most efficient to create a set of small command files, as below, and to use them when needed (e.g. for analyses in Chapters 10 and 12), rather than adding them into the data definition file. Some analyses use the original computed variables, and some use the variables after they have been recoded into categories. The relevant information is provided as the examples are presented in the text.

EXTRANEU recodes personality items E1 to E12 and computes a neuroticism score (*neur*) and an extraversion score (*extra*).

```
recode e1 to e12 (2 = 0) (3 = -1).
compute neur = (e1 + e3 + e5 + e6 + e9 + e11).
compute extra = (e2 + e4 + e7 + e8 + e10 + e12).
```

RECEN recodes *neur* and *extra* scores into three categories (with approximately equal numbers in each).

```
recode extra neur (lo thru 0 = 1) (1 thru 3 = 2) (4 thru hi = 3).
```

SEXISM computes a *sexism* score from 7 of the 10 attitude items *var001* to *var010*.

```
compute sexism = (var001 + var002 + var009 + (8 - var003) +
(8 - var004) + (8 - var005) + (8 - var008)).
```

RECSEX   recodes *sexism* into three categories (with approximately equal numbers in each).

```
recode sexism (14 thru 20 = 1) (21 thru 26 = 2) (27 thru 38 = 3).
```

SEXTRACT   computes a *sextract* score from 6 of the 10 attitude items *var001* to *var010* (using the same formula as for *sexism* but omitting reference to *var003*).

```
compute sextract = (var001 + var002 + var009 + (8 - var004) +
(8 - var005) + (8 - var008)).
```

RECSEXCT   recodes *sextract* into three categories (with approximately equal numbers in each).

```
recode sextract (10 thru 18 = 1) (19 thru 23 = 2) (24 thru 35 = 3).
```

PROBLEM   computes a *problem* score from 7 of the 10 personal problem items *prs01* to *prs10*.

```
compute problem = (2 - prs01) + (2 - prs03) + (2 - prs04) +
(2 - prs07) + (2 - prs08) + (2 - prs09) + (2 - prs10).
```

Appendix G: Glossary of Statistical, Computing and SPSS/PC+ Terms

SPSS/PC+ keywords and commands (including subcommands) are printed in **bold type**.

ABSOLUTE VALUE: the value of a number or expression which ignores whether the original value is positive or negative. Thus the absolute value of -34 (and also of $+34$) is 34.

ACTIVE FILE: an SPSS/PC+ file, with data and dictionary information, which is currently in use. Certain changes (transformations, selections, etc.) may have been made since the file was originally made active and these changes may remain in effect until the end of the session. Thus the term active file refers to the file in use in its current state.

ADVANCED STATISTICS: the name of an optional add-on enhancement for SPSS/PC+. It includes a number of advanced methods of analysis, including factor analysis, cluster analysis, loglinear modelling, discriminant analysis and multivariate analysis of variance. The use of this package and these techniques are not described in this book. However, the methods of data entry, file creation, transformation, etc. applicable to the Advanced Statistics enhancement are the same as those used for the SPSS/PC+ base system.

ALL: an SPSS/PC+ keyword which can sometimes be used, instead of a list, to request that all variables be analyzed, or all statistics performed.

ALPHANUMERIC VARIABLE: (also referred to as a string variable) a variable which includes letters or symbols. Because the values are in character (rather than number) form, only a limited number of operations (including sorting) can be performed with alphanumeric variables. They cannot be subjected to arithmetic operations. *See also* LONG STRING; SHORT STRING.

ANALYSIS OF VARIANCE: an inferential statistical procedure which allows comparison of the means (for a dependent variable) of groups defined by the values of one or more independent variables. If a single independent variable is involved then the SPSS/PC+ procedure **ONEWAY** may be used. If more than one independent variable is involved then

ANOVA is the appropriate procedure. *See* Chapter 12.

AND: a logical operator used to join two or more relations, for example in a conditional transformation. Thus 'IF (VAR1 EQ 2 AND VAR2 EQ 3) ...'.

ANOVA: an SPSS/PC+ command which performs analysis of variance.

AREA SAMPLING: a method of sampling in which an area is taken and divided into a number of smaller areas, or blocks. A sample of these blocks is then randomly selected, and samples are taken only from the selected blocks.

ARITHMETIC OPERATORS: the signs which give instructions for arithmetic operations to be carried out. The four basic arithmetic operations are addition ($+$), subtraction ($-$), multiplication ($*$) and division ($/$). SPSS/PC+ also allows exponentiation to be directly specified with '$**$'. Thus '$2**3$' would be read as: '2 to the power of 3', or '2 cubed' (i.e. 8). With these basic operations (and with brackets) complex transformations may be specified (for example, in a COMPUTE command). *See also* FUNCTIONS.

ASCII: The American Standard Code for Information Interchange – a standard code which converts alphabetic and numeric information, and certain symbols and control characters into binary code. Information coded according to this standard, and in which lines are terminated by a <RETURN> and the file is terminated with a <Control Z>, are text files and can be written and read by a number of editors and by SPSS/PC+.

AUTOEXEC.BAT: a DOS batch file which is *auto*matically *exe*cuted when DOS is loaded (or booted). Appendix A gives instructions for creating such a file for use with SPSS/PC+.

BACKUP: to make a copy of files on a separate disk. It is always advisable to keep a spare diskette copy of files in case the hard disk copies are accidentally destroyed.

BARCHART: an SPSS/PC+ subcommand (of the FREQUENCIES command) which requests that the frequencies of data be plotted as a bar chart.

BAR CHART: a diagramatic representation of data in which frequencies for a variable with relatively few distinct values are plotted graphically. *See also* HISTOGRAM. ·

BATCH FILE: *See* BATCH MODE.

BATCH MODE: a system whereby data or instructions are first prepared and then processed as a batch by the system. SPSS/PC+ command files which have been prepared using an editor are batch files. Information given to the system in response to the SPSS/PC: prompt, however, is said to be entered in interactive mode. In DOS a batch file is a file which contains one or more DOS commands. It is commonly given the file extension .BAT. Sometimes a batch file is created which will be *auto*matically *exe*cuted when DOS is loaded (or booted). Such a batch file is named "autoexec.bat".

BEGIN DATA: an SPSS/PC+ command which is included within a command file or data definition file containing data. The 'begin data.' command (including the period, **.**) is entered immediately before the first line of data and before any procedures have been specified. A 'begin data.' command is *not* used when data are read from a separate data file. *See also* END DATA.

BEEP: the sound issued by the system to signal an error (low-pitched sound) or the availability of the next screen of output (high-pitched sound). The beep can be switched off by the SET command: set beep off.

BINARY CODE: a system for coding information which uses only the digits 0 and 1.

BINARY SITUATION: a situation (e.g. experiment, test, or question) in which there are only two possible outcomes (e.g. the toss of a coin, or a question to which the only permitted answers are 'yes' or 'no').

BINOMIAL: an SPSS/PC+ subcommand (of the NPAR command) which requests a binomial test. *See* Chapter 13.

BUFFER: a small part of the computer's memory which acts as a temporary information storage space. For SPSS/PC+ to run efficiently the number of available buffers should be set to at least 8. If a suitable 'CONFIG.SYS' file has been written, this will be done automatically when the machine is switched on. *See* Appendix A.

BY: an SPSS/PC+ keyword. Used, for example, in a CROSSTABS command, e.g 'crosstabs sex BY soclass.'

CASE: in a survey, a case would usually represent one respondent. In a psychological experiment a case might represent one experimental subject (i.e. one person). A dataset generally consists of the values relating to a standard set of variables for a number of cases. A case need not represent an individual person – it might represent a country, corporation, product, etc.

$CASENUM: a system variable which is automatically included for each case in the active file. The $ symbol at the beginning of the name indicates that the variable is a system variable rather than a user-declared variable. The value of $CASENUM is the position of that case in the dataset as the file is made active. If the order of cases is later changed by a SORT command, or if the size of the active file is reduced, the value of $CAS-ENUM remains unchanged. Thus the original order of the initial active file (or what remains of it) can be re-established with the command: 'SORT by $CASENUM.' If the active file is saved as a system file, the $CASENUM assigned when the file was made active is included within the system file. $CASENUM is not, however, included in a portable file created with EXPORT.

CATEGORICAL VARIABLE: a variable for which all values conform to particular categories. Only a limited range of arithmetic procedures are permitted with such variables. *See also* CODE.

CELL: one intersection in a crosstabulation. *See also* CHI-SQUARED TEST.

CENTRAL TENDENCY (MEASURES OF): the collective name for various descriptive statistics which identify different kinds of mid-point in a frequency distribution. Commonly, the mean, median, and mode.

CHARACTER: a letter, digit from 0 to 9, or any other recognized symbol or punctuation mark. The standard character set used in most computing is the ASCII set.

CHISQUARE: an SPSS/PC+ subcommand (of the NPAR command) which requests a 'One sample chi-squared' test. *See* Chapter 13. Note also that the chi-squared test can be requested as a STATISTIC on the CROSS-TABS command.

CHI-SQUARED TEST: a statistical test which determines whether an observed pattern of results differs significantly from that which would be expected to occur merely by chance. Thus if it were assumed that by chance an equal number of boys and of girls would have blue eyes (rather than brown eyes) we could count the number of boys and girls with the two types of eye colour and subject the results to a chi-squared test. A *significant* result would indicate that for our sample sex and eye colour were related – that the distribution of eye colour between the sexes was *not* random (or the result of mere chance). This would be an example of a 2×2 chi-squared (here, two sexes and two eye colours). The frequencies for each sex and each eye colour could be displayed in a cross-tabulation with four cells.

CLOSED-ENDED QUESTION: a question for which a number of fixed alternative answers are provided. Examples included seven-point rating scales, Likert scales and other forms of multiple choice. By limiting valid responses to a fixed range of answers the data obtained is easily codable and readily amenable to SPSS/PC+ analysis.

COCHRAN: an SPSS/PC+ subcommand (of the NPAR command) which requests a Cochran test. *See* Chapter 13.

CODE: a category label. To code data is to categorize it. Some category systems reflect an underlying dimension – 'tall', 'average height', 'short' – and some do not – 'blue-eyed', 'brown-eyed'. The labels may be numbers (1 for 'blue-eyed', 2 for 'brown-eyed') but this should not be taken to imply that the codes express ranks or quantities. The term coding data is often used to describe the process of creating the complete data matrix from a study, a process that might involve codes and real numbers.

CODEBOOK: a handy list, written by the user in the early stages of preparing a project, which contains the names of the variables to be included in a file, with information about the location, values, missing value, coding system, and labels for each variable. *See* Appendix D for a full example of a codebook.

CODING: the process of translating raw data into categorized data.

CODING SYSTEM: any system by which raw data are categorized for analysis. Many coding systems are simple (for example, we could code people's ages into three categories – under 20; 20 to 40; over 40). Some are more complex – marital status, for example, may be coded using categories which include a consideration of marital history ('twice widowed', 'twice married') as well as to present status. Very elaborate coding systems may need to be devised (usually after the data has been collected) if open-ended questions have been used.

COMMAND: an instruction issued to the SPSS/PC+ system either in batch mode or interactively. Operation commands display or change the current operating characteristics of the system. Data definition and transformation commands instruct the system on where and how data should be read, how data should be transformed and selected, and how output should be labelled. Procedure commands cause the data values to be read and, usually, instruct the system to perform a statistical analysis. Procedure commands may include one or more subcommands. Commands begin with a command word (or a three-character truncation of the command word) and must end with a period (.).

COMMAND FILE: an SPSS/PC+ file which contains a number of commands regarding procedures, transformations, selections, etc. This file is submitted to the SPSS/PC+ system as a batch file (using the command 'include'). A data definition file is one example of a command file, and the names of such files commonly include the extension .DEF. The names of other command files often include the extension .SPS.

COMPUTE: an SPSS/PC+ command which creates a new variable from the existing dataset (or modifies the values of an existing variable). It takes the form: COMPUTE C = A + B. Complex COMPUTE commands are possible and may include several arithmetic operators, functions, relational operators and logical operators. The creation (or modification) of each variable needs a separate compute command.

CONDITIONAL TRANSFORMATION: an SPSS/PC+ data transformation which depends on one or more logical conditions being fulfilled. *See* LOGICAL EXPRESSION; *see also* Chapter 8.

CONFIG.SYS: a DOS file which sets certain parameters to configure or customize the system for a particular type of use. Instructions contained within a CONFIG.SYS file are automatically executed when DOS is loaded (or booted). Appendix A gives instructions for creating such a file for use with SPSS/PC+.

CONTINGENCY TABLE: a tabulation of the joint frequencies for two or more variables. *See* CROSSTABULATION.

CONTINUOUS VARIABLE: a numerical variable for which the values need not be integers. Thus height can be represented with decimal point data (e.g. 163.54) whereas an integer variable like 'number of children' must be a whole number (1, 2, 3, etc.). Much of the data to be analyzed by SPSS/PC+ will be coded in number form but will be categorical rather than numerical data. Because they are not numerical, coded variables should not be treated as either continuous variables or as true integer variables.

CORRELATION: an SPSS/PC+ command which causes a product-moment correlation to be performed.

CORRELATION: the correspondence between two sets of paired measurements. The numerical index used to indicate the degree of this correspondence is the correlation coefficient. The most common measure is the product-moment correlation (symbolized by r) based on a formula devised by Karl Pearson. The value of a correlation coefficient ranges from -1 (indicating total negative correlation) through 0 (indicating no association between the two sets of data) to $+1$ (indicating complete correspondence). A high positive correlation does not imply that the mean (i.e. average) values for each of the two variables are equal or similar but implies that high values of one variable are associated with high values of the other. A multiple correlation is a measure of the degree of relationship between a single dependent variable and a number of independent variables. *See also* CORRELATION MATRIX, PARTIAL CORRELATION.

CORRELATION MATRIX: a table of results of correlational analyses in which the correlation coefficients between each possible pair of a number of sets of data are represented.

CROSSTABS: an SPSS/PC+ procedure which produces a crosstabulation. The command takes the form: 'crosstabs sex by soclass.'; values of the first named variable determine the rows of the table, and values of the second named variable determine the columns. Various OPTIONS affect the display and the treatment of missing data, and various STATISTICS are available, including chi-square.

CROSSTABULATION: (or contingency table) a table showing the joint distribution of frequencies of cases on two or more discrete variables. Thus the frequency distribution for two sexes and three hair colours will be represented in a crosstabulation containing six cells. Crosstabulation is requested by using the SPSS/PC+ CROSSTABS command.

CURSOR: a lighted square or line on the computer screen which marks the current position at which information can be written.

CURVILINEAR RELATIONSHIP: a relationship between two variables which is shown graphically not as a straight line but as a curve.

DATA: a collection of numbers, measurements or codings.

DATABASE: an organized collection of data. *See also* DATABASE MANAGEMENT SYSTEM.

DATABASE MANAGEMENT SYSTEM: a program which organizes and manipulates data to produce summaries, comparisons, lists, etc.

DATA DEFINITION FILE: a command file which includes the data dictionary, i.e. information about the variables, their format, variable and value labels, missing data information, etc. The only essential command within a data definition file is the DATA LIST command which declares variables to the system. Some data definition files also include the data itself (more often they will contain a reference to a data file).

DATA ENTRY: an optional enhancement of the SPSS/PC+ system which provides an easy way of entering data. The valid range for each variable can be pre-set, so that any attempt to enter a datum which is outside the valid (or missing) range will be recognized by the system, and an error message given.

DATA FILE: a text file which contains only data.

DATA LIST: an SPSS/PC+ data definition command which specifies the names of variables and their position in the dataset. The data list may refer to in-line data or specify the name of a data file from which the data are to be read. In the latter case the form of the command is: DATA LIST FILE = "gender.dat" (followed by the list of variables and their locations). By default, fixed format is assumed, but free format can be specified by: DATA LIST FREE (or, where a data file is to be read: DATA LIST FILE = "gender.dat" FREE (followed by the list of variables). A DATA LIST command may also refer to matrix data.

DATASET: a set of data. Usually comprises (in the case of SPSS/PC+ datasets) data for a number of variables from each of a number of cases.

DATUM: the singular of data; one item within the dataset.

DEFAULT: a value, option, name or instruction which is assumed by the system when no specification has been provided by the user. A passive default comes into operation without any action on the part of the user (and most defaults are of this type). An active default, however, requires the user to request that a default should be brought into operation. Thus if a STATISTICS subcommand is included within a FREQUENCIES procedure without further specification, four default statistics are displayed. With no STATISTICS subcommand, however, no statistics are displayed.

DEGREES OF FREEDOM: a statistical term which refers to the number of values within a dataset which are free to vary. For example, if we know that the sum of the ages of six children is 40 then it is clear that knowing the ages of any five of the children will enable us to calculate the age of the sixth child. Thus the degrees of freedom in such a case would be five. When there is a single restriction (in this example, the combined age) the general formula for calculating the degrees of freedom is $N - 1$ (in this case $6 - 1 = 5$). Degrees of freedom is included as an element in many statistical formulae and its value in any particular case will affect the value of the relevant statistic. The degrees of freedom will also need to be taken into account in assessing the significance of this value. SPSS/PC+ assesses degrees of freedom automatically, using it both in its calculation of any relevant statistic (such as the t-value) and in its calculation of the probability of the statistic. The value of the degrees of freedom is also displayed in the output from many procedures.

DEMOGRAPHIC: demography is the study of (usually human) populations. Demographic data includes information on such variables as age, sex, income, social class and marital status. Such information is often collected in surveys on attitudes, etc. so that analyses relevant to these major population characteristics can be performed. This is the case even when it is not the major purpose of the survey to focus on such parameters.

DEPENDENT VARIABLE: a variable which is measured in two or more conditions. In an experiment, an independent variable (for example, level of noise) is manipulated and a dependent variable (for example, accuracy of word recognition) is measured. The term is also applied in other types of study. Thus in an attitude survey the sex of the respondent may be treated as an independent variable and a particular attitude as a dependent variable.

DESCRIPTIVE STATISTICS: an area of statistical analysis which describes data (their frequencies, averages, spread, correlations, etc.). This contrasts with inferential statistics.

DICTIONARY: information, within a SPSS/PC+ file, about the variables referred to in that file. The dictionary can contain, besides the name of the variable, indications about its format and location within the dataset, the missing value, and both variable labels and value labels.

DIRECTORY: a logical (rather than physical) part of a hard disk. Because a hard disk may store many hundreds of files, some logical structure is necessary to allow efficient access by the user. There is a basic root directory and many sub-directories (often one for each of the major users). From within any one directory, access to other directories can be gained by the use of a path command. Consult the user manual for the DOS system for further information.

DISCRETE VARIABLE: a variable which can take one of only a restricted set of values from within the overall range. The most common form of discrete variable is the integer variable in which only whole numbers can be appropriately used. Thus any answer to the question 'How many children in the family?' must be a whole number (0, 1, 2, 3, etc.). This variable is thus a discrete (or integer) variable.

DISK DRIVE: a part of the computer hardware. Its function is to read information from and write information to a diskette or hard disk. *See* drive A.

DISKETTE: a flexible disk (usually 5.25 inches in diameter) which can be used in drive A of IBM microcomputers and compatible machines. A diskette is also sometimes referred to as a floppy disk.

DISPLAY: an SPSS/PC+ operation command which lists the variables currently defined to the system. The command DISPLAY, without further specification, lists the names of the variables and any labels which have been provided for them. If the command specifies one or more variables by name (e.g. display sex age var001) then additional information (format, missing value and value labels) is displayed for each variable specified.

DOS: abbreviation for Disk Operating System. The set of programs which manages file manipulation, directory organization and the interaction between the computer processor and the monitor, keyboard, printer, etc. SPSS/PC+ uses MS-DOS.

DRIVE A: one of the computer's disk drives; a part of the computer hardware which can read information from and write information to a disk. Drive A refers to a drive which deals with diskettes, (including the SPSS/PC+ key diskette and those introduced by the user to copy information to and from the hard disk). If a second diskette drive is present it will be drive B. Whether one or two diskette drives are included, drive C will refer to the hard disk. The SPSS/PC+ system is resident on the hard disk unit and it is this drive that will be the most used during the SPSS/PC+ session.

DRIVE B: *see* drive A

DRIVE C: *see* drive A

DROP: an SPSS/PC+ subcommand that can be used with the commands IMPORT and EXPORT, SAVE and GET, and JOIN. It is followed by a list of variables and instructs the system to exclude the named variables when executing the particular operation. Thus with an active file of 100 variables, if the SAVE command includes a DROP subcommand listing 10 of these variables, the saved system file will contain only 90 variables. When DROP is used, the variables in the new file remain in the same order as they were in the original file. *See also* KEEP.

ECHO: normally, commands issued to the system by the user are shown on the screen but are not printed and are not included within the log file. The SET command: SET ECHO ON., however, will cause any commands to be echoed on the printer. They will also be included in the log file.

EDITOR: a program which allows text to be written and edited. SPSS/PC+ has its own editor, REVIEW, but many other editors, including word-processing programs, database management systems, etc. can be used to create and edit files for use by SPSS/PC+.

ELSE : a keyword which can be used within a RECODE command, e.g.: RECODE VAR1 (0,1,2 = 1) (3,4,5 = 2) (ELSE = 3). ELSE recodes **all** other values including user missing and system missing values.

END DATA: an SPSS/PC+ command which is included within a command file or data definition file containing in-line data. The END DATA. command is entered immediately after the last line of data and before any procedures have been specified. Unlike most keywords the END DATA. cannot be truncated but must be written in full. The command is *not* used when data are read from a separate data file. *See also* BEGIN DATA.

EQ: a relational operator meaning equal to. Thus IF (VAR1 EQ 3) means 'if the value of VAR1 is equal to 3'. The equals symbol (=) can be used to replace EQ.

ERROR: a problem diagnosed by the system. When the system diagnoses a mistake or inconsistency it will display an error message and will not execute a procedure. Amongst the most common errors are the incorrect spelling of a command word, blank lines inserted within a multi-line command, and a terminator period omitted from the end of a command. Mistakes in the declaration of variable and value labels, however, and most mistakes in data entry, will not be identified by the system.

EXPORT: an SPSS/PC+ command which produces a portable ASCII file from the current active file. Portable files include both data and data dictionary information as well as certain system information. They can be transported to other computers using Kermit and are imported into the SPSS system of the other machine using the IMPORT command.

FACTOR ANALYSIS: a complex statistical technique which computes, from a correlation matrix, the minimum number of 'factors' which will account for the overall pattern of intercorrelations. It provides a synopsis of the information contained in such a pattern. Factor analysis is used, for example, to identify the basic factors being measured by a large number of items within a personality inventory and may also be used to derive a suitable scoring method for such an instrument. Factor analysis is a procedure included in the SPSS/PC+ *Advanced Statistics* enhancement, and it is not treated in this volume.

FILE: a collection of information. Within the SPSS/PC+ system there are many types of file. They vary in terms of their content, their purpose and the form in which information is stored. Thus, in terms of content, some SPSS/PC+ files contain only data, others include data definition information, and some contain SPSS/PC+ commands. Others contain results of analyses or a record of commands entered during a session. For some types of file, information is stored in ASCII form. System files, however, are stored in binary form. The texts of some files are written by the user (using an editor). Other files are written by the system at the request of the user, and some are written automatically by the system. Appendix H gives a table of the characteristics of the different types of file. *See also*: ACTIVE FILE; COMMAND FILE; DATA FILE; DATA DEFINITION FILE; LISTING FILE; LOG FILE; PORTABLE FILE; RESULTS FILE; SYSTEM FILE.

FILE NAME: the name of a file, consisting of up to eight characters and an extension of three letters following a period. Most file names are specified by the user, but for some files the system provides default names (e.g. "SPSS.LIS" for the listing file and "SPSS.LOG" for the log file). Although most file names do not **need** to include an extension, it is often useful to provide this. For example, .DEF may be used as the extension for a data definition file, .DAT for a data file, .SPS for a command file, and .MAT for a matrix results file.

FINISH: an SPSS/PC+ command which terminates the current SPSS/PC+ session and returns the user to DOS.

FIXED FORMAT: data is commonly organized within the dataset by placing data for the same variable for each case within precisely the same column(s), thus forming a perfectly rectangular data matrix. The system can then be informed of the position occupied by each variable. Fixed format is the default format used by SPSS/PC+ and it contrasts with free format in which successive variables are separated by one or more blank spaces.

FLOPPY DISK: a diskette (usually 5.25 inches in diameter) which can be used in drive A of IBM microcomputers and compatible machines.

FORMAT:
(a) the form, and position within the database, of data for each of the variables. A format specification for the data relating to a particular variable in an SPSS/PC+ file may include information about the location of the variable (i.e. columns occupied), the position of an implied decimal point (*see* Chapter 16) and the numeric or alphanumeric nature of the variable. *See also*: FIXED FORMAT; FREE FORMAT.
(b) to format a disk is to prepare a new disk so that it will be compatible with the computer system being used. Thus new diskettes to be used in drive A will have to be formatted appropriately before files can be written to them.
(c) generally, the term format describes the position and layout of a display or item. SPSS/PC+ allows the user a high level of control over format, often by means of a FORMAT subcommand.

FORMAT: a subcommand used within a number of procedure commands to specify how lists, plots, and output displays should be printed or displayed or how a results file should be written.

FORMATS: an SPSS/PC+ command which allows the user to change the way in which numeric variables are displayed and printed. The formats specified on such a command remain in effect for the rest of the session unless a new FORMATS specification is provided.

FREE FORMAT: a way of writing a data line, in which successive variables are separated by one or more blank spaces (and/or commas). This contrasts with the more usual fixed format method (the SPSS/PC+ default). When free format is used there is no need to add leading zeros to even up the space occupied by variables (as when 003 is entered for 3) and thus the data for different cases may be of different lengths. There is no insistence on a perfectly rectangular matrix for the dataset as there is for datasets of fixed format.

FREQUENCIES: an SPSS/PC+ procedure used to produce frequency counts for each value of each variable specified, together with percentages. Subcommands can also be used to request various formats for the output, including barcharts and histograms. Many optional statistics are also available, and these must be requested by name (*see* list in Chapter 6) rather than by number.

FREQUENCY CURVE: a plot of the distribution of values for a variable. A graphical representation of the frequency distribution.

FREQUENCY DISTRIBUTION: a set of scores in which data are grouped into classes and the frequency obtained for each class.

FRIEDMAN: an SPSS/PC+ subcommand (of the NPAR command) which requests a Friedman test. *See* Chapter 13.

FUNCTIONS: most commonly, transformations of numbers (e.g. square root, log). Functions can be used within SPSS/PC+ COMPUTE, IF, and SELECT IF commands. A large range of functions can be employed within SPSS/PC+, including numeric functions (e.g. ABS for absolute value, SQRT for square root), missing value functions (e.g. SYSMIS), etc. The full range of functions is given in the COMPUTE command section of the *SPSS/PC+ Manual*.

FUNCTION KEYS: a set of special keys on the computer keyboard. When the SPSS/PC+ editor program REVIEW is in operation, several of these have special assigned effects. These effects differ depending on whether the function key is pressed alone, with <SHIFT>, with <ALT>, or with <CONTROL>. A guide to these functions is provided on-screen (when REVIEW is in operation) by pressing <F1> (alone).

GE: a relational operator meaning greater than or equal to. Thus IF (VAR1 GE 3) means 'if the value of VAR1 is greater than or equal to 3'.

GET: an SPSS/PC+ command which reads a system file previously written to disk with a SAVE command. If no filename is provided the file with the default name "SPSS.SYS" is read.

GRAPHICS: an optional enhancement of the SPSS/PC+ system which produces high-quality pictorial representations of features of the data. Among the many available display formats are pie charts, bar charts, line graphs and scatterplots. GRAPHICS provides the user with an extensive range of options for controlling the size, shape and shading of graphs and, with a suitable printer, the system can produce colour graphics.

GT: a relational operator meaning greater than. Thus IF (VAR1 GT 3) means 'if the value of VAR1 is greater than 3'.

HARD COPY: a printed copy of the contents of a file. It is useful to have a hard copy of all data files and definition files.

HARD DISK: technically, a Winchester disk unit. A rigid disk which stores information magnetically. Most hard disks store at least 10 megabytes of information – equivalent to that contained in 30 novels.

HARDWARE: The physical components of the computer system, including the keyboard, monitor, etc.

HELP: a command issued interactively during a session to obtain on-line help about a topic. Help is available for general topics (e.g. HELP transformations.), for commands (e.g. HELP crosstabs.) and for subcommands (e.g. HELP crosstabs options.).

HI (or **HIGHEST**): a keyword which can be used within a RECODE command, e.g. RECODE VAR1 (1 THRU 3 = 1) (4,5 = 2) (6 THRU HI = 3). HI should be used with care – if the highest value, or a value within a range specified by HI, is a user-missing value this too is recoded.

HISTOGRAM: an SPSS/PC+ subcommand (of the FREQUENCIES command) which requests that the frequencies of data be plotted as a histogram.

HISTOGRAM: a diagramatic representation of data in which frequencies for different value-ranges of a variable are plotted graphically. Thus to plot a histogram of age, for a relatively small population with a large age range, we would first have to group data into value-ranges (under 20, 20 to 30, etc.). Where there are only a few distinct values for a variable (for example, social class coded into five categories), such grouping of values is unnecessary and a bar chart may be used.

IF: an SPSS/PC+ command which specifies that a transformation (the creation of a new variable or a change in values for an existing variable) should be made IF certain conditions are fulfilled. *See* Chapter 8.

IMPORT: an SPSS/PC+ command which reads a portable ASCII file. Portable files include both data and data dictionary information as well as certain system variables and system information. They can be transported from another computer using Kermit and are then imported into the system when specified on an IMPORT command.

INCLUDE: an SPSS/PC+ operation command used to read a command file.

INDEPENDENT VARIABLE: a variable having two or more conditions in which a dependent variable is measured. Thus in an experiment, an independent variable (sometimes called the experimental variable – say, the number of words in a list) is manipulated and a dependent variable (for example, the proportion of words remembered) is measured. The term is also applied in other types of study. Thus in an attitude survey the social class of the respondent may be treated as an independent variable. The attitudes examined in association with social class would be dependent variables.

INFERENTIAL STATISTICS: an area of statistical analysis which permits conclusions (inferences) to be drawn about whether a characteristic is significantly different in two or more populations, etc. This type of analysis contrasts with descriptive statistics which merely describe data (their frequencies, averages, spread, correlations, etc.). *See also* SIGNIFICANCE TESTS.

INITIALIZATION:
(a) (also referred to as re-initialization) the process by which files with the names "SPSS.LIS" and "SPSS.LOG" are erased at the beginning of an SPSS/PC+ session. These are the default names for the listing file and the log file. To keep a more permanent record of a session the user should provide different names for the listing and/or log files using the SET command.
(b) the process of assigning a common value for a new variable to all cases before specifying (by means of a COMPUTE or an IF command) that certain cases should take specific values. Thus COMPUTE TEST = 0 would assign the value 0 to the new variable *TEST* for all cases. A subsequent command: IF (ESSAY1 GE 50) TEST = TEST + 1 would increase the value of *TEST* to 1 for all cases with an *ESSAY1* value of 50 or above). Without the initialization command, the value for *TEST* for those cases with an *ESSAY1* value of less than 50 would be system missing.

IN-LINE DATA: data included within a data definition file or a command file, or entered interactively during a session. A BEGIN DATA. command must be placed immediately before the first line of data, and an END DATA. command placed on a line immediately after the last line of data.

It is more usual for data to be entered into a separate data file and for this file to be read into the system when it is specified on a data list command included in a command file.

INPUT: to enter information into a computer.

INSTALLATION: the process of copying the diskettes on which SPSS/PC+ is originally supplied onto the hard disk and setting certain parameters of the computer system to make it suitable for SPSS/PC+ analysis. *See* Appendix A for details.

INTEGER VARIABLE: a variable which can take only whole number values. Any variable which is a count is an integer variable. Thus the number of children in a family must be 1 or 2 or 3, etc. and cannot be 1.56 or 2.35.

INTERACTION EFFECT: the extra combined effect of two or more independent variables on a dependent variable after the main effects have been taken into account. The term is used, for example, in two-way analysis of variance designs.

INTERACTIVE MODE: *see* interactive operation.

INTERACTIVE OPERATION: the issuing of commands directly to the machine in response to the SPSS/PC+ prompt. This mode of operation contrasts with the alternative mode (known as batch processing) in which a sequence of commands is written into a command file and this file then included.

INTER-QUARTILE RANGE: *see* quartile.

INTERVAL SCALE: a scale of measurement in which the data are truly quantitative (unlike nominal or ordinal data) and for which arithmetic operations such as addition, subtraction, multiplication and division are permissible. For an interval scale, equal **differences** between two numbers (e.g. the difference between 3 and 5, and between 10 and 12) signify equal differences in the quantity being measured. Unlike a ratio scale, however, the interval scale does not imply that there is a fixed zero-point. Parametric statistics can be used with interval data and ratio data. A general discussion about scales of measurement will be found at the beginning of Chapter 13.

JOIN: an SPSS/PC+ command which allows the user to combine two, three, four or five files (one of which can be the active file – the others must be system files) to become a new active file. JOIN MATCH effectively adds two (or more) sets of variables for the same (or overlapping) cases. JOIN ADD adds cases (with overlapping variables) to form a combined population. *See* Chapter 15.

K-S: an SPSS/PC+ subcommand (of the NPAR command) which requests a Kolmogorov-Smirnov test. *See* Chapter 13.

K-W: an SPSS/PC+ subcommand (of the NPAR command) which requests a Kruskal-Wallis test. *See* Chapter 13.

KEEP: an SPSS/PC+ subcommand used with the commands IMPORT and EXPORT, and JOIN. It instructs the system to include only specific variables when executing the operation. Thus if, with an original portable file of 100 variables, the IMPORT command includes a KEEP subcommand listing only 65 of these variables, the resulting active file will contain only those 65 variables. When KEEP is used, the variables in the new file are placed in the order in which they are specified by the KEEP subcommand. *See also* DROP.

KENDALL: an SPSS/PC+ subcommand (of the NPAR command) which requests a test for Kendall's coefficient of concordance (*W*). *See* Chapter 13.

KERMIT: a file transfer program which allows SPSS portable files (including files originally created on the SPSS/PC+, SPSS/PC or SPSS^X systems) to be transferred between computers. Compatible versions of Kermit must be available on *both* of the machines. Kermit is an error-correcting program which monitors and checks the information-transfer process.

KEY DISKETTE: a special diskette (or floppy disk) which must be placed in drive A in order for SPSS/PC+ to operate. The key diskette provides a way of copy-protecting the valuable system and thus prevents unauthorized use. If you copy files from the key diskette, you may find that the SPSS/PC+ system will no longer load.

KEYBOARD: the typewriter-like part of the computer system which allows the user to input information. It includes the alphabetic keys and a number of additional keys, including a number pad, the 'control' key, the 'alt' key, and 'function keys'.

KEYWORD: a word which has a special assigned meaning in SPSS/PC+. As well as command and subcommand words, certain other words, including BY, WITH, TO, ALL, AND, THRU, ELSE and NOT, are keywords. It is important to avoid keywords when naming variables.

KURTOSIS: the extent to which the data peaks around the middle value or is flat across the range of frequencies. The standard error of the kurtosis is a measure of the confidence we can have that the kurtosis of the current sample is a true representation of the kurtosis of a larger theoretical population.

LABEL: an optional string of characters used to provide annotations for SPSS/PC+ output. Variable labels (specified on a VARIABLE LABELS command) are used to provide an extended description of variables. Value labels (specified on a VALUE LABELS command) allow the user to provide brief descriptions of the particular values which a variable may take. Labels must always be enclosed in apostrophes (or quotation marks) when specified in commands. Although labels may be up to 60 characters in length it is best to keep them as succinct as possible. All labels are optional. If a spelling error is included in a label specification it will have no effect on the operation of the program (but, of course, the labelling within the output will include the erroneous spelling).

LE: a relational operator meaning less than or equal to. Thus IF (VAR1 LE 3) means 'if the value of VAR1 is less than or equal to 3'.

LEAST SQUARES METHOD: a method of determining the form of curve (of a specified kind) which best fits a set of data.

LEVELS OF MEASUREMENT: *see* scales of measurement.

LIKERT SCALE: a three- or five-point scale in which the respondent indicates the degree to which s/he agrees with a statement. The points on the scale are often labelled 'strongly agree', 'agree', etc.

LISTING FILE: a text file which is automatically written by SPSS/PC+ and contains any output which has appeared on the screen (or in printed form) as a result of procedures activated during the session (including warnings and error messages, and also page numbers and titles). The default name for the listing file is "SPSS.LIS". This file is initialized when the next SPSS/PC+ session begins. Listing files can be printed or edited, or can be copied to a diskette for off-site editing.

LISTWISE DELETION: an SPSS/PC+ term describing one way in which missing data are excluded from analyses. If a particular case has missing data for any variable included in a list of variables, then the case is excluded from all procedures and calculations specified for that list. This strategy contrasts with that of pairwise deletion.

LO (or **LOWEST**): a keyword which can be used within a RECODE command, e.g.: RECODE VAR1 (LO THRU 2 = 1) (3,4,5 = 2) (6,7 = 3). LO should be used with care – if the lowest value, or a value within a range specified by **LO**, is a user missing value or the system missing value this too is recoded.

LOCATION: the position within the dataset of the data for a particular variable. In fixed format, data for a particular variable must be located in the same column(s) for all cases.

LOG FILE: a text file which is automatically written by SPSS/PC+ and contains any commands entered during the SPSS/PC+ session. The log file also contains information that warnings or error messages were issued, and provides references to output page numbers. The default name for the log file is "SPSS.LOG". This file is initialized (overwritten) when the next SPSS/PC+ session begins. Log files can be printed, edited or copied to a diskette for off-site editing.

LOGICAL EXPRESSION: an expression which is evaluated for its truth value. It is frequently encountered in a conditional transformation. Thus 'IF (A = B) C = 1.' Here A = B is the logical expression. If, for a particular case, the expression is TRUE (i.e. the value of A **does** equal the value of B), then, for that case, C will be set to equal 1. Logical expressions may be very complex and may include many relational operators (equals, less than, etc.). Two or more expressions may be joined by one or more logical operators.

LOGICAL OPERATOR: an operator used in conditional transformations to join two or more logical expressions. SPSS/PC+ employs the logical operators NOT, AND and OR. *See* Chapter 8.

LONG STRING: a string variable from 9 to 255 characters long. Each segment of 8 characters counts as one variable towards the system limit of 200 variables. *See also* STRING VARIABLE; SHORT STRING.

LT: a relational operator meaning less than. Thus IF (VAR1 LT 3) means 'if the value of var1 is less than 3'. The less than symbol (<) can be used to replace LT.

MAIN EFFECT: the effect of a single independent variable on a dependent variable (for example, in an analysis of variance). *See also* INTERACTION EFFECT.

MAP: an SPSS/PC+ subcommand used within the commands IMPORT and EXPORT, and JOIN. It produces a listing of variable names from the current active file dictionary. When used with a JOIN command, MAP also specifies the original files from which each of the variables in the composite file has come. With EXPORT and IMPORT, MAP lists the actions implemented up to that point (thus allowing the user to keep track of variables which have been dropped, kept and re-named). MAP can be used several times within a single command; if it is placed as the last subcommand then it provides a list of the final contents of the active file.

MATRIX: a two-dimensional table of data values or results. For some SPSS/PC+ procedures (e.g. CORRELATION, ONEWAY) matrices may be written to a results file by specifying a particular OPTION. Such a matrix may then be used as input to another procedure, increasing the efficiency and speed of the analysis. Data that is entered into a procedure in this way is referred to as matrix data.

MATRIX DATA: some SPSS/PC+ procedures (e.g. ONEWAY and REGRESSION) will accept summary statistics (i.e. the results of previous analyses of the relevant data) as their input data. If data is to be entered in this form a MATRIX specification must be used on the DATA LIST command.

MAXIMUM: the highest value for a particular variable within a set of data.

MCNEMAR: an SPSS/PC+ subcommand (of the NPAR command) which requests a McNemar test. *See* Chapter 13.

MEAN: the arithmetic average. The sum of all scores divided by their number. The standard error of the mean is a measure of the confidence we can have that the average based on the present sample is a true representation of the average of a larger theoretical population.

MEDIAN: an SPSS/PC+ subcommand (of the NPAR command) which requests a 'Median' test. *See* Chapter 13.

MEDIAN: the midscore. The value which is a midpoint, dividing the population into 50 per cent below the median and 50 per cent above the median.

MINIMUM: the lowest value for a particular variable within a set of data.

MISSING DATA: if, for a particular case, the value for a variable is unknown (for example, if a person has not answered a question on a questionnaire) then the relevant data is missing. Such circumstances are represented in the dataset by a missing value which has been assigned for that variable by the user. This is the user-missing value. In certain circumstances a missing value is assigned to a variable by the system. For example, when a new variable is created by a COMPUTE command, any case for which the variable cannot be computed is assigned a missing value for that variable. This is the system-missing value, and it is indicated in screen displays and printed output by a period.

MISSING VALUE: an SPSS/PC+ data definition command that provides a value (or code) which, when encountered in the data for that variable, is to be interpreted by the system as indicating that the real value (for this variable and for this case) is unknown (i.e. missing). Data having the missing value are ignored during statistical calculations.

MODE: the most frequent value for a variable.

MODULE: the unit for the removal and installation of SPSS/PC+ procedures. The various SPSS/PC+ procedures (both those included in the base system and those included in the optional enhancements) are grouped together into modules. Some of these are non-permanent and can be removed from the system and re-installed using the SPSS MANAGER procedure. *See* SPSS MANAGER, and Appendix A.

MONITOR: the visual display unit. The screen displays both the input to the system and the output from the system. SPSS/PC+ can also send output to a printer.

MORE: often, it will not be possible for all of the information to be provided by SPSS/PC+ (results, help information, etc.) to be accommodated on a single screen. The first screenful of information will be displayed and the operation then paused to allow the user time to read the contents. At the same time the message MORE will be displayed. When any key is pressed, the next screenful of information will be displayed. Sometimes the results of a single analysis will be presented over as many as twenty displays. If the user wishes to read the results from printed output, rather than from the screen, the MORE function serves no purpose. It may be switched off by entering the SET command 'set more off.' in response to the SPSS/PC+ prompt.

MOSES: an SPSS/PC+ subcommand (of the NPAR command) which requests a Moses test. *See* Chapter 13.

MS-DOS: the disk operating system (*see* DOS) used by SPSS/PC+. This set of programs controls the interaction between different hardware elements within the system, including the operation of peripherals (printer, screen, disk drives, etc.), and controls such functions as the copying, deleting, and printing of files.

MULTIPLE COMPARISON TESTS: statistical tests used (for example, after a one-way analysis of variance – using the SPSS/PC+ procedure ONEWAY) to determine which pairs of groups are significantly different. Thus if an analysis of variance has shown that there are overall differences in height between five samples from different ethnic groups, multiple comparison tests will show which particular groups differ significantly from which others. The multiple comparison tests performed after a general effect has been identified are sometimes referred to as *post-hoc* tests. In the ONEWAY procedure seven multiple comparison tests are available (Duncan's multiple range test, the Tukey test, etc.). Each must be requested using a separate RANGES subcommand.

MULTIVARIATE ANALYSIS: the name given to any statistical technique which attempts to identify the component structure of complex data.

Examples of such analyses include factor analysis, cluster analysis and loglinear analysis (all of which are included in the optional SPSS/PC+ *Advanced Statistics* enhancement).

M-W: an SPSS/PC+ subcommand (of the NPAR command) which requests a Mann–Whitney *U* test. *See* Chapter 13.

N (command): an SPSS/PC+ selection command which places a limit on the size of the active file to the first N cases. The command takes the form N 54. *See* Chapter 8 for further information.

NE: a relational operator meaning not equal to. Thus IF (VAR1 NE 3) means 'if the value of VAR1 is not equal to 3'.

NOMINAL SCALE: a scale of measurement in which data are grouped into name classes between which there are no formal mathematical relationships. Thus although classes are different from each other there is no implied order of the categories. Although categories may be assigned numerical codings (e.g. 1 for male and 2 for female) such numbers act only as labels and should not be used in calculation. Certain non-parametric statistics including the chi-squared test, the binomial test and the Cochran test, are suitable for the analysis of nominal data.

NON-PARAMETRIC STATISTICS: statistical methods which do not assume that data are distributed normally (*see* NORMAL DISTRIBU-TION). Frequently used non-parametric methods include the chi-squared test and other tests included in the SPSS/PC+ **NPAR** procedure. *See* Chapter 13.

NORMAL CURVE: also known as the Gaussian curve or the probability curve. A plot of the normal frequency distribution. *See* NORMAL DIS-TRIBUTION.

NORMAL DISTRIBUTION: a frequency distribution which conforms to a standard bell-shape. The three measures of central tendency – the mode, mean and median – are at the same point, and the distribution is symmetrical. If a set of data conforms (or nearly conforms) to this distribution then statistical inferences can be made about it using parametric statistics.

NOT: a logical operator used, usually in a conditional transformation, to reverse the true–false status of the logical expression which follows it. Thus IF (NOT VAR1 EQ 2) GROUP = 1.

NPAR: an SPSS/PC+ procedure for the non-parametric analysis of data. **NPAR** includes a number of different tests for the analysis of categorical and ordinal data from one or more groups. *See* Chapter 13.

NULL HYPOTHESIS: the hypothesis that any statistical difference or association obtained will be the result of mere chance. If the result of an analysis falls outside the range which can reasonably be attributed to chance fluctuations then that result is said to be statistically significant and the null hypothesis is therefore rejected. The conclusion in such a study would be that the effects are *not* merely the result of chance. The indirect strategy of rejecting the null hypothesis is linked to assumptions underlying statistical procedures.

NUMERICAL EXPRESSION: an expression which consists of numbers or numeric variables linked by operators such as +, − and /. Thus (VAR001 + 6 − VAR005) is a numerical expression. Such an expression might be used, for example, in creating a new variable with a **COMPUTE** command.

NUMERIC VARIABLE: a variable for which the values are represented by true numbers (rather than codes). Data for such variables can therefore be used legitimately in arithmetic calculations.

OFF-SITE: the term used in this book to refer to the creation or editing of a file on a microcomputer other than that on which SPSS/PC+ is resident. If the SPSS/PC+ machine is used by several people, the most efficient way of sharing the resource may be for data files and data definition files to be prepared on another machine, and for priority in the use of the SPSS/PC+ facility to be given to those who wish to execute SPSS/PC+ procedures.

ONE-TAIL TEST: a test for the significance of a difference in a case in which the hypothesis has been cast in a directional form. Contrasts with the two-tail test of a non-directional hypothesis. *See* TAIL; SIGNIFICANCE TESTS.

ONEWAY: an SPSS/PC+ statistical procedure used to perform a one-way analysis of variance.

ON-LINE: immediately accessible. It is usual for computer programs to run one at a time, but sometimes one program, or a different part of the same program, may be called into use while the other is still active. Where such a situation occurs the second program or the second part of the same program is said to be available on-line − it is immediately accessible and will not interfere with the program currently in operation. The SPSS/PC+ **HELP** facility (a part of the SPSS/PC+ program) is available on-line in this way. It gives on-screen information even when another part of the program (e.g. **REVIEW**) is in use.

OPEN-ENDED QUESTION: a question which permits the respondent to answer in free form, not limiting the response to a choice between given alternatives. Examples: 'Where were you born?'; 'Name three things you like' and 'Give any ideas you have for improving health care.' Such questions often provide rich information, but the responses may be difficult to analyze. A coding system may be devised to classify or score the responses. Open-ended questions contrast with closed-ended questions.

OPERATOR: a symbol denoting a mathematical or logical process. Operators are used in various SPSS/PC+ commands, and especially in conditional transformations and selections. We can distinguish between arithmetic operators (+, /, etc), relational operators (**EQ, NE, GT,** etc.) and logical operators (**AND, OR** and **NOT**).

OPTIONS: a subcommand available with many procedures which allows the user to specify various features of the analysis and output. The **OPTIONS** subcommand often allows the user to specify how missing data should be treated, the form the analysis should take, and how the output should be formatted and labelled. For some procedures, the **OPTIONS** subcommand is also used to specify that input data should be read from a matrix, or that results should be written to a results file in matrix form.

OR: a logical operator used to join two or more relations in a conditional transformation. Thus IF (VAR1 EQ 2 OR VAR1 EQ 3) ... and SELECT IF (VAR1 EQ 2 OR VAR1 EQ 3). Notice that such a statement is written in a full form . The short form IF (VAR1 EQ 2 OR 3) would **not** be valid.

ORDINAL SCALE: a scale of measurement in which the numbers refer to positions in an ordered series. Thus in a race, the results in terms of first, second, third, etc. may be represented by 1, 2, 3, etc. but these are ranks and should not be regarded as real numbers which can be added, subtracted, multiplied, etc. If two or more items share the same placing then there are said to be 'tied ranks'. Certain non-parametric statistics, including the Moses test and the Wilcoxon test, are suitable for use with ordinal data.

ORTHOGONAL: forming right angles. If there is no association between two variables or two factors resulting from a factor analysis (i.e. if the correlation between them is 0) they are described as orthogonal.

OUTPUT FILE: a file created by SPSS/PC+ at the request of the user. The commands which lead to the creation of output files are **WRITE** (which produces a rectangular results file), **SAVE** (which creates a system file) and **EXPORT** (which creates a portable file).

PAIRWISE DELETION: an SPSS/PC+ term describing one way in which missing data are excluded from analyses. If a particular case has missing data for one or both of two variables specified for a particular procedure that case is excluded from the calculation. Otherwise it is included. This strategy contrasts with that of listwise deletion.

PARAMETER: any measurable characteristic of a population.

PARAMETRIC STATISTICS: statistical methods which make certain assumptions about the parameters of the population from which the sample has been drawn. In particular, they assume that observations are independent and that measurements are continuous. They may also assume that the data frequencies are normally distributed (or approximate to such a distribution). Statistics such as the *t*-test, product-moment correlation and analysis of variance depend on the data having such characteristics, although in practice such tests are able to return meaningful results even with data for which certain of the assumptions do not hold.

PARTIAL CORRELATION: the correlation between two variables which remains when the effects of certain other variables have been partialled out (i.e. eliminated or corrected for).

PERCENTILE: a range which contains a hundredth part of an ordered dataset. If the data has been arranged in a frequency distribution of ascending values then the value which cuts off the lowest valued hundredth part of the set is the first percentile. The fiftieth percentile will correspond to the median. If the sample is normally distributed then percentile values can be calculated from a knowledge of the mean and the standard deviation.

PLOT: an SPSS/PC+ procedure which produces a graph of the frequencies of data for two variables simultaneously. *See* Chapter 10.

POISSON DISTRIBUTION: a highly asymmetric frequency distribution which is obtained when very rare random events are plotted.

POPULATION: a complete set of objects, individuals or measurements which have some characteristic in common. A group from which a sample may be drawn.

POOLED VARIANCE: *See*: SEPARATE VARIANCE.

PORTABLE FILE: an ASCII file produced by SPSS/PC+, SPSS/PC or SPSS^X, using the **EXPORT** command. Portable files include a copy of the data, data definition information, system variables and system information. The file transfer program Kermit is used to transfer portable files between mainframe computers running SPSS^X and microcomputers running SPSS/PC+. Portable files are not easily editable. *See* Chapter 16.

PROCEDURE: a type of SPSS/PC+ command, often a request for a statistical analysis, which reads data values from the active file. If no active file is current, the procedure command will create such a file by following any 'data read' commands which have previously been issued. Before beginning the requested analysis, procedure commands execute any transformations which have been requested but not yet put into effect.

PROCESS IF: a command which selects cases to be included in the next procedure if certain conditions are fulfilled. The conditions are specified by a logical expression which may include relational operators (such as NE – not equal to or LT – less than) and logical operators. Thus: PROCESS IF (VAR1 NE 5 OR VAR2 LT 7). **PROCESS IF** leads to a selection for the **next procedure only**. It does not lead to cases being discarded from the active file, whereas the command **SELECT IF** permanently reduces the number of cases in the active file. *See also*: SELECTION.

PRODUCT-MOMENT CORRELATION: a test designed by Karl Pearson to determine whether high values of one variable are associated with high scores on another (and low values on one with low values on the other). The calculation produces the statistic r, the correlation coefficient, which varies between -1 (a perfect negative correlation) to $+1$ (a perfect positive correlation). This statistic is calculated by the SPSS/PC+ **CORRELATION** procedure.

PROMPT: an instruction which appears on the screen to give information and invite a response from the user. A number of common screen prompts are used by SPSS/PC+. When the system is at rest waiting for a command, the SPSS/PC+ prompt SPSS/PC: is displayed.

QUARTILE: one of three values which divide a frequency distribution into equal quarters. These values will depend on the range and the shape of the frequency distribution. The first quartile marks the 25th percentile, the second marks the 50th percentile and the third marks the 75th percentile. The distance between the first and third quartiles (therefore spanning the range containing the central 50% of the distribution) is known as the inter-quartile range.

QUOTA SAMPLING: a sampling method by which respondents are selected to be representative of a larger population with respect to certain (usually demographic) characteristics such as age, social class and sex. Thus if it were known that 10% of all surgeons in the total population being studied were female, a quota sample would ensure that 10% of the sample was female.

RANDOM: occurring by chance alone. Random error is that part of the variability of a measure which is due to chance, and such error is found to be normally distributed around the mean.

RANDOM SAMPLE: a selection chosen at random from a population and assumed to be representative of it.

RANGE: the difference between the lowest value and the highest value for a variable.

RANGES: a subcommand used (within the ONEWAY procedure) to request one of seven statistical multiple comparison tests. The request for each of these tests must be specified on a separate **RANGES** subcommand.

RANK: the position of an item in an ordered series.

RATING SCALE: a means by which a respondent can record his or her judgement according to a numerical or analogue system with well-defined end-points. Thus a scale for judging pain might involve the end-point descriptions 'no pain' and 'intense pain'. Gradings might be made on such a scale in terms of a fixed number of points (7-point scale, 11-point scale, etc.). Alternatively, the respondent might be asked to mark an appropriate point on a continuous line to indicate the degree of pain involved. The response would then be scored by measuring the distance of the respondent's mark from the end of the line.

RATIO SCALE: a scale of measurement in which data are truly quantitative. Ratio data share with interval data the fact that equal **differences** between numbers (e.g. the difference between 3 and 5, and between 10 and 12) are equal. For ratio data, however, the zero point is not arbitrary but represents the true absence of the quality being measured. Height in centimetres, for example, is a ratio measurement. A ratio scale represents the highest level of measurement.

RAW DATA: the data as they are originally obtained, i.e. before they have been subjected to any transformation, grouping or manipulation.

RECODE: an SPSS/PC+ command which is used to revise the values of an existing variable. Thus if the user needs to compare an old group with a young group, and *age* is currently coded in terms of years, the following **RECODE** could be used to produce just two age-groups: RECODE age (10 thru 30 = 1) (31 thru 98 = 2). *See* Chapter 8.

REGRESSION: linear regression is a relationship between two variables when their values, plotted graphically, produce a straight line. Multiple regression is the relationship between a single criterion variable and a number of predictor variables.

REGRESSION: an SPSS/PC+ procedure for performing linear and multiple regression. It is a complex procedure and details are not included in this book.

RELATIONAL OPERATOR: a symbol that is used in a logical expression to compare one value with another. Thus one value may be equal to (**EQ**) another, not equal to (**NE**), greater than (**GT**), or less than (**LT**) another, etc. Relational operators are used in SPSS/PC+ conditional transformations (**IF, SELECT IF** and **PROCESS IF**). *See* Chapter 8.

RELIABILITY: the extent to which a test has internal consistency. Test–retest reliability, which is usually expressed in terms of a correlation coefficient, is an assessment of the similarity between two measurements of a variable obtained by applying the test on two separate occasions.

REPEATED MEASURES DESIGN: an experimental design in which an individual or group is tested repeatedly after exposure to a number of experimental conditions. The **MANOVA** procedure from the *Advanced Statistics* enhancement is available for the analysis of data from such research designs.

REPORT: an SPSS/PC+ procedure used to produce tables, including listings of cases and summary statistics. Before a **REPORT** command is issued the data must be organized into an appropriate order using a **SORT** command. Data can be broken down by break variables (using a **BREAK** subcommand) and the display is then specified by one or more **SUMMARY** subcommands. Various labels may also be specified. *See* Chapter 14.

RESPONDENT: usually, a person who completes or answers a questionnaire. More generally, any case in the sample that has been surveyed or observed.

RESPONSE RATE: the number of returned questionnaires as a proportion of those sent out.

RESULTS FILE: a file which includes results from a specific procedure. It can be produced using the **WRITE** command or by specifying an appropriate **OPTION** with certain procedures (for example **CORRELATION** and **ONEWAY**). By default, the results file is named "SPSS.PRC". This can be changed using the **SET** command. Although results files *can* be edited with an editor any changes made in this way may make them unsuitable as input to other procedures. Editing of these files is therefore generally not recommended. If the user needs an editable file which includes results of procedures, the listing file will generally prove more suitable.

REVIEW: the SPSS/PC+ text editor which allows the editing of text files (including data files, data definition files, command files, log files and listing files). **REVIEW** can be entered directly from DOS. If it is entered during an SPSS/PC+ session, however, the command **REVIEW** (with no file name specified) will produce two windows on the screen. The top window will show the most recent part of the listing file and the bottom part will show the most recent part of the log file. The cursor can be moved from one window to the other and can be used to scroll to the appropriate part of the text. If the command is entered with a file name (e.g. **REVIEW** "gender.dat") the appropriate file is made available for editing. An on-screen guide to the various **REVIEW** functions can be obtained from within REVIEW by pressing <F1>.

RUNS (TEST): a non-parametric test, also known as the Wald–Wolfowitz test, available within the **NPAR** procedure. *See* Chapter 13.

SAMPLE: a selection of scores from a larger (often theoretical) population. If the sample is random then it is assumed to be representative of the larger population.

SAMPLE: an SPSS/PC+ command which draws a temporary sample of cases from those currently in the active file. The usual form is '**SAMPLE** .33'. This would lead to approximately one third of cases being selected.

SAMPLING ERROR: the variation in scores or statistics which derives from the non-representativeness of the sample.

SAMPLING PROCEDURES: strategies for selecting a sample effectively and efficiently so that valid conclusions can be drawn regarding a larger population. Various sampling methods are used, including area sampling, quota sampling, and stratified sampling.

SAVE: a command which produces an SPSS/PC+ system file. The default name for a file created with **SAVE** is "SPSS.SYS" but an alternative name can be specified (e.g. **SAVE OUTFILE**= "gender.sys"). There is also an opportunity, when saving a system file, to **DROP** some of the variables included in the current active file.

SCALE: an instrument for assigning a value to an item to give a measure on some dimension.

SCALES OF MEASUREMENT: also referred to as levels of measurement. Numbers are used in different ways, and have different properties, depending on how they are used. Some numbers merely act as labels, some refer to ranks, and some represent true measures. A number which is a label, for example, cannot be added – it would be foolish to think that two football players numbered 1 and 3 were somehow equivalent to one player labelled 4. Different types of statistical analysis are appropriate when numbers are used in these different ways. Scales of measurement refers to these different ways of using numbers. The most commonly distinguished scales are nominal, ordinal, interval, and ratio.

SCATTER PLOT: a graph in which items are plotted according to their values on two variables, one of which is scaled on the horizontal axis and the other of which is scaled on the vertical axis.

SCROLLING: the process of moving information up and down the screen in order to get to the part that needs to be examined or worked on.

SELECT IF: a command which selects cases to remain in the active file if certain conditions are fulfilled. The conditions are specified by a logical expression which may include relational operators (such as **EQ** – equal to or **GT** – greater than) and logical operators. Thus SELECT IF (VAR1 GT 3 OR VAR2 EQ 8). **SELECT IF** permanently affects the active file whereas the command **PROCESS IF** specifies which cases should be included in the next procedure (only). *See also* SELECTION.

SELECTION: the process of including only certain cases for analysis or manipulation (transformation, etc.). The two SPSS/PC+ commands **SELECT IF** and **PROCESS IF** may be used to select subsets of the original population according to criteria specified by the user. For example SELECT IF (SEX = 1). might lead to only the male cases remaining in the active file.

SESSION: an SPSS/PC+ session begins when the 'spsspc' command is issued and the SPSS/PC+ logo is displayed on the screen, and it lasts until the command **FINISH** is issued in response to the SPSS/PC+ prompt. Within a session many different files can be made active, and many different analyses performed. If an active file is tailored (for example with a **SELECT IF** command, or a **RECODE** command) then the changes remain in effect for the rest of the session (unless the original file is made active again). Log files and listing files are maintained for each session and are overwritten at the beginning of the next session (unless their names have been changed from the default names).

SET: an SPSS/PC+ command which allows the user to specify a number of options affecting the way in which the system behaves, e.g. the names to be given to the listing file and log file and the on/off condition of the printer, the MORE feature and the BEEP. Thus to switch the printer on, the command would be: SET printer on.

SHORT STRING: a string variable up to eight characters in length. Such strings can be used in transformation commands and **SORT** commands but are not available for arithmetic operations. *See also* STRING VARI- ABLE; LONG STRING.

SHOW: an SPSS/PC+ command which produces a screen display of infor- mation about the current state of certain parameters concerning the run- ning of the system. These include the memory space available, the current names for the log file and listing file, etc. Many of these parameters can be changed by an appropriate use of the **SET** command.

SIGNIFICANCE TESTS: some statistical procedures merely describe or summarize data – these are descriptive statistics. Others examine the significance of a difference or an association between samples – these are inferential statistics. When a difference or association has been found we need to establish whether the magnitude of the effect is within the range which could reasonably be explained by mere chance factors or whether it is significant i.e. unlikely to be a chance effect. One common criterion for accepting an effect as significant is that an effect of such a magnitude would occur by chance less than 1 in 20 times. If the result of the statistical procedure reaches this level it is said to be significant at the 5% level. A result which would occur less than 1 in 100 times by chance is significant at the 1% level (etc.). Traditionally the researcher, having calculated a value for the statistic (say *t* or chi-squared) would consult a set of statistical tables to determine whether the value was significant. SPSS/PC+ automatically provides information about the significance of many of the statistics calculated. A statistically significant value may or may not be significant in other ways, and the validity of a statistical inference will always depend on the appropriateness of the test which has been used. It is possible, particularly with packages such as SPSS/PC+, to obtain complete statistical analyses – including estimates of statistical significance – on data which completely violate the assumptions which must be fulfilled in order for the test to be validly applied.

SIGN (TEST): a non-parametric test, available within the **NPAR** procedure, used to determine whether there is a significant difference between matched pairs of data. The test takes into consideration only the direction of the differences between pairs of data and ignores the magnitude of such differences. *See* Chapter 13.

SIGNED-RANKS TEST: a non-parametric test, also known as the Wilcoxon, available within the **NPAR** procedure. *See* Chapter 13.

SKEWNESS: a measure of the lop-sidedness of a frequency distribution – i.e. a measure of the tendency for data to cluster towards the lower or upper end of the range rather than being evenly distributed or balanced around the midpoint. The standard error of the skewness is a measure of the confidence we can have that the skewness of the particular sample is a true representation of the skewness of a larger theoretical population.

SOFTWARE: computer programs. The SPSS/PC+ system is a highly complex and sophisticated program (software) run on the computer machinery (the hardware).

SORT: a procedure command which re-arranges cases into a particular order within the active file according to one or more key variables (such as *age* or *surname*). The **SORT** command allows cases to be sorted alphabetically or numerically and in either ascending or descending order. Ascending, here, means low to high (or, for an alphabetic sort, A to Z).

SPSS.LIS: the default name for the listing file. This file is initialized at the beginning of each SPSS/PC+ session. The name of the listing file can be changed at any time by issuing a suitable **SET** command. The listing from that point to the end of the session will then be stored on a file with the new name, and this will not be inititialized at the beginning of the next session.

SPSS.LOG: the default name for a log file. This file is initialized at the beginning of each SPSS/PC+ session. The name of the log file can be changed at any time by issuing a suitable **SET** command. The log from that point to the end of the session will then be stored on a file with the new name, and this will not be inititialized at the beginning of the next session.

SPSS MANAGER: a procedure used to remove SPSS/PC+ modules and to reinstall them to the system.

SPSS.PRC: the default name for a results file. This name can be changed by a suitable **SET** command, e.g.: SET RESULTS = "ATT.PRC".

SPSSPROF.INI: the file name given to a batch file containing commands which should be issued to the system each time SPSS/PC+ is entered from the particular directory. These commands often include SET commands which affect the operation of various parameters of the system (printer, beep, more and echo on/off status, for example, and log and listing file names). Once an "SPSSPROF.INI" file has been written and saved within a particular directory, the commands will be automatically executed each time the system is run from within that directory.

SPSS.SYS: the default name for a system file. A file of this name will be created if no file-name specification is given on a **SAVE** command, and this file will be read by a subsequent **GET** command which has no file-name specification. Only one "SPSS.SYS" file can be present in any particular hard disk directory, so such a SAVE command will cause any existing file of that name to be overwritten by the new file. The default name can be changed by a suitable **SET** command.

STANDARD DEVIATION: a measure of the spread or scatter of values around the mean. If the frequency of values for a variable conforms to a normal distribution then knowledge of the mean and standard deviation allows an estimate to be made of the percentage of cases whose values fall below or above any particular value for that variable.

STANDARD ERROR (S.E.): generally, a measure of the degree to which a statistic is likely to differ from its true value as a result of chance factors. For explanations of particular types of standard error (e.g. s.e. of skewness) refer to the entry for the relevant measure (e.g. skewness).

STANDARD SCORE: also known as the Z-score, this is a transformation of the raw score which takes into account the mean and standard deviation of the population. The standardized score is the difference between the raw score and the population mean, divided by the standard deviation. Standard scores have a mean of 0 and a standard deviation of 1, although other transformations may be applied at a secondary stage.

STATISTICS: an SPSS/PC+ subcommand available for many procedures which allows the user to specify which statistical tests (in addition to the default tests for that procedure) should be performed.

STRATIFIED RANDOM SAMPLE: *See* STRATIFIED SAMPLING.

STRATIFIED SAMPLING: a method of sampling in which a population is first divided into sub-groups, or strata, according to some characteristic, and a random sample then taken of each sub-group.

STRING VARIABLE: (also known as an alphanumeric variable). A string is a sequence of characters. A string variable is one which includes letters or symbols. Because such variables are not numeric, only a limited number of operations (including sorting with the **SORT** command) can be performed with them, and they are not available for arithmetic operations. *See also* LONG STRING; SHORT STRING.

SUBCOMMAND: an instruction which forms part of a command and specifies which operations are to be performed or the form in which output is to be displayed. For some commands, only, one or more subcommands are necessary. The most familiar subcommands are those which specify **STATISTICS** and **OPTIONS**.

SUM: the result of adding each of the values for a particular variable in a set of data.

SURVEY: a technique for gathering data from or about a group of people or objects. Such methods include behavioural observation, written questionnaires and telephone or face-to-face interviews. In a sample survey the aim is to use a relatively small group (the sample) in order to derive conclusions or predictions about a larger group (the population). The extent to which valid extrapolations to the larger population will be possible will depend on the representativeness of the sample. *See* SAMPLING PROCEDURES.

SYSMIS: a special SPSS/PC+ function used in COMPUTE commands and conditional transformations and selections. The function produces the value 1 (or true) if the named variable has a system missing value. Thus: SELECT IF (SYSMIS(VAR101)). would retain in the active file only those cases for which *VAR101* had a system-missing value.

SYSTEM FILE: a file created from an active file by using the SPSS/PC+ **SAVE** command. System files contain data, data definition information, system variables and other system information, and they are written in binary form. The use of system files can lead to considerable reduction in processing time, but they have the disadvantage that they cannot be edited.

SYSTEM-MISSING VALUE: a missing value automatically assigned by the SPSS/PC+ system in certain circumstances. Thus if a **COMPUTE** command specifies that C is to be created for each case by adding the value of A to the value of B, then if one or both of these values are missing from the original dataset then C, for that case, will be assigned a system-missing value. *See also* **sysmis**.

SYSTEM VARIABLE: the SPSS/PC+ system automatically assigns values to certain special variables for its own house-keeping. These are known as system variables. One such variable is *$CASENUM* (the names of system variables always begin with the $ symbol). Cases are assigned a case number by the system as a file is made active. One way in which the user may employ this variable is to use it as a sorting key (within a **SORT** command) to restore the original order of cases within the active file: SORT BY $casenum. *See also* $CASENUM; $WEIGHT.

TABLES : an optional enhancement of the SPSS/PC+ system which produces high quality summary tables. **TABLES** provides the user with great flexibility in determining the appearance and content of tables.

TABULATION: an arrangement of information in the form of a table. *See also* CROSSTABULATION; CONTINGENCY TABLE.

TAIL: either extreme of a frequency distribution. The term is most commonly applied to probability distributions. A hypothesis about a difference between two sets of data can be stated in a directional form (e.g. boys will be taller than girls) or in a non-directional form (e.g. there will be a difference, one way or the other, in the heights of boys and girls). When a *t*-test is used to determine whether the means of two groups differ significantly the calculation will result in a particular *t*-value. To assess the significance of this value the directional or non-directional nature of the hypothesis must be taken into account. If the hypothesis is directional then a one-tail test applies. If the hypothesis is non-directional then a two-tail test applies.

TEST: a collection of items which can be scored together to yield one or several measurements of individual differences (e.g. personality, levels of ability).

TEXT FILE: a file containing alphanumeric characters (and possibly some control characters). Text files are generally accessible for editing. Some files used by SPSS/PC+ are of this type although system files are not.

THRU: an SPSS/PC+ keyword used within the **RECODE** command to help specify a range of values to be recoded to a single new value. For example: RECODE VAR1 (1 THRU 99=1) (100 THRU 199=2). The range specified by THRU includes both of the limiting values, so that the expression '1 THRU 5' includes 1 and 5, and all values within this range.

TITLE: a command used to provide a page heading for SPSS/PC+ output. The default heading includes the date, the page number and 'SPSS/PC+'. If a TITLE has been specified then SPSS/PC+ is replaced with the user-specified title (up to 58 characters long). Thus: TITLE "Gender Questionnaire - Student Data".

TO: an SPSS/PC+ keyword used to list variables on a **DATA LIST** or **PROCEDURE** command. Thus VAR1 TO VAR44 would list 44 variables (*VAR1, VAR2, VAR3 ... VAR43, VAR44*). TO can also be used within the **LIST** command to refer to cases. Thus LIST CASES FROM 1 TO 58 will produce a display of the values of all of the variables for cases 1 to 58.

TOGGLE: a switching between two states or modes of operation. For example, pressing <Ins> repeatedly will cause the system to change between the insert and overwrite modes in **REVIEW**. This action thus operates the insert ON/OFF toggle.

TRANSFORMATION: in SPSS/PC+, the act of changing the coding scheme for a variable (with a **RECODE** command) or of creating a new variable (with a **COMPUTE** command). SPSS/PC+ has powerful facilities for transforming data. *See also* CONDITIONAL TRANSFORMATION, Chapter 8.

TRUNCATION: an SPSS/PC+ feature which allows the user the enter a keyword (e.g. **CORRELATION, VARIABLES**) in a shortened form. Only the first three letters are needed (e.g **COR, VAR**). Because of this feature, mispellings **after** the first three letters are ignored – thus **CORELASHUN** is an acceptable form of **CORRERLATION**.

t-TEST: a parametric statistical test used to determine whether the means of two groups are significantly different. There are two versions of this test. The paired-samples *t*-test (also known as the matched pairs *t*-test, the related *t*-test, and the correlated *t*-test) compares the means of two paired sets of data. The two-sample *t*-test (also known as the independent *t*-test) compares the means of two unrelated or independent samples. *See* Chapter 7.

T-TEST: an SPSS/PC+ command used to request either a paired-samples or an independent-samples *t*-test. *See* Chapter 7.

TWO-TAIL TEST: a test for the significance of a difference between the means of two groups in a case in which the hypothesis has been cast in a non-directional form. This contrasts with a one-tail test. *See* TAIL; SIGNIFICANCE TESTS.

TYPE I ERROR: the conclusion that there is a difference between two populations when, in fact, there is **not**.

TYPE II ERROR: the conclusion that two populations are **not** different when in fact they **are**.

USER-MISSING VALUE: a value for a particular variable which has been declared by the user (in a **MISSING VALUE** command) as representing data which are missing. The system recognizes the special nature of this value and, unless specifically requested to do so, will generally not treat the value literally. Thus if the value 9 has been declared as signifying missing data (e.g. for a variable for which only the values 1 to 8 are valid) any 9's encountered for this variable will not be included in calculations of averages, etc.). *See also* SYSTEM-MISSING VALUE.

VALID: in SPSS/PC+, any value for a variable other than that which has been declared by the user as a missing value and that which is a system-missing value.

VALIDITY: generally, the extent to which something meets a criterion of reasonableness or accuracy. Of a test or measure, the degree to which it measures what it is supposed to measure. There are several different ways of assessing the validity of a test. Typically, a set of scores provided by the test is compared with some external criterion, for example the performance in a task. A valid sample is one which has been appropriately selected and is therefore assumed to be representative of the larger population.

VALUE: a number or code which represents, for a particular case, the characteristic (size, degree, type, etc.) of the variable. The term is generally applied to quantified variables, but can also refer to those which have been coded into categories. The value in such cases refers to the code – a category **label** rather than a true number.

VALUE LABEL: an optional description of some or all of the particular values which a variable may take, e.g. (for *sex*) 1, 'male' 2, 'female'. The output from many SPSS/PC+ procedures will include any relevant value labels which have been provided by the user. Value labels must always be enclosed in quotation marks when specified in a command. Although they may be up to 60 characters in length, it is preferable to keep them as succinct as possible. Value labels may be included within a data definition file or provided interactively during an SPSS/PC+ session.

VARIABILITY: the scatter of values in a sample.

VARIABLE: an item which can assume one of several different values or characteristics.

VARIABLE DEFINITION: that part of the **DATA LIST** command which declares the names of the variables represented by the data and specifies the format and position of each variable.

VARIABLE LABEL: a description of the variable which can be provided by the user, and is used by SPSS/PC+ to annotate output. Variable labels should always be enclosed in quotation marks when specified in commands, and although they may be up to 60 characters in length it is preferable to keep them succinct. Variable labels may be included within a data definition file or provided interactively during an SPSS/PC+ session.

VARIANCE: a measure of the dispersion of values. The square of the standard deviation.

WALD–WOLFOWITZ (TEST): a non-parametric test, also known as the Runs test, available within the **NPAR** procedure and used to determine whether a sequence of data with only two possible values (for example heads and tails) is random or contains 'runs' of one or other value. *See* Chapter 13.

WARNING: the response made by the system to alert the user to a possible mistake. If a long string variable, for example, is specified in a command which applies only to numerical and short string variables then a warning is issued. The procedure continues, however, with the long string truncated to the first eight characters (i.e. the system treats it as a short string). *See also* ERROR.

$WEIGHT: a system variable which is automatically included for each case in the active file. The $ symbol at the beginning of the name indicates that the variable is a system variable rather than a user-declared variable. $WEIGHT is an extra variable for each case which represents a weighting to be applied to the case in analyses. The initial value of **$WEIGHT** for each case is 1.00. This can be changed by a **WEIGHT** command (e.g. WEIGHT BY VAR004.). The differential weighting can be cancelled with the command: WEIGHT OFF.

WILCOXON (TEST): a non-parametric test, also known as the signed-ranks test, available within the **NPAR** procedure. It is used to determine whether matched pairs of data from two groups (at at least the ordinal level of measurement) show a significant difference. *See* Chapter 13.

WILDCARD CHARACTER: a character which, when employed in certain DOS operations involving file-names, is interpreted to mean any single character or, alternatively, any group of characters. When using DOS to copy files, the name of each file to be copied can be specified in full. Alternatively, the wildcard characters ? and ★ can be employed. Thus the copying of three files named "genone.dat", "gentwo.def" and "gen-three.com" could be requested using a single command which includes the formulation copy gen?????.★. Here each question mark would refer to any character and the asterisk would refer to 'any group of characters'. The special formula '★.★' is used to copy **all** files (whatever their name and extension) from one drive to another. When wildcard characters are used to copy a group of files each file retains its original name when copied to the second drive.

WINDOW: when the monitor screen is divided into two or more logically separate sections, each is known as a window. For example, issuing the **REVIEW** command during an SPSS/PC+ session (without specifying the name of a file to be reviewed) leads the editor program **REVIEW** to display two windows. The top window reveals a part of the listing file and the bottom window reveals a part of the log file .

WRITE: an SPSS/PC+ procedure which creates a results file of cases together with variable names and format information. Details of the use of the **WRITE** command are not provided in this book. Results files are also written if the user specifies certain **OPTIONS** within a number of procedure commands (including **CORRELATION** and **ONEWAY**).

X-AXIS: the horizontal axis (abscissa) of a graph.

Y-AXIS: the vertical axis (ordinate) of a graph.

Z SCORE: the standard score.

Appendix H: Types of SPSS/PC+ Files

File type	Includes	Written by USER(U) or SYSTEM(S)	Write command	Default name	Read command	Type (EXT.)	Edit?
Data	Data	U	(Editor)	None	DATA LIST	ASCII (.DAT)	Yes
Data definition	Data List, labels, etc. (and can also include data)	U	(Editor)	None	INCLUDE	ASCII (.DEF)	Yes
Command	Commands (and can also include data definition info. and data)	U	(Editor)	None	INCLUDE	ASCII (.SPS)	Yes
System	Data and definition information	S	SAVE	SPSS.SYS	GET	Binary (.SYS)	No
Listing	Output from any procedures during current session	S	None (auto)	SPSS.LIS	REVIEW	ASCII (.LIS)	Yes
Log	A record of input during current session	S	None (auto)	SPSS. LOG	REVIEW	ASCII (.LOG)	Yes
Portable	As system files (+some system information)	S	EXPORT	None	IMPORT	ASCII (.POR)	Not easily
Results	Matrix of summary results	S	Approp. OPTION selected	SPSS.PRC	INCLUDE (or DATA LIST)	ASCII (.PRC)	Yes

Appendix I: SPSS/PC+, SPSS/PC and SPSS^X COMPARED

The version of SPSS currently used on mainframe computers is known as SPSS^X. The first 'micro' version of SPSS was known as SPSS/PC, and this has now been superseded by the current micro version, SPSS/PC+.

In most ways the different versions overlap, but there are some notable differences. The purpose of this Appendix is to provide an **overall** profile of the kinds of differences between these versions, rather than to give detailed information. This will be of use both to those who have read this book, and now wish to use one of the other versions of SPSS, or to those who are already familiar with one of the other versions and now wish to explore the use of SPSS/PC+. Full details of the differences are provided in Appendix G of the *SPSS/PC+ Manual*.

I.1 SPSS/PC+ and SPSS^X

Depending on the particular characteristics of the mainframe implementation of SPSS, the user may find the experience of using both mainframe and microcomputer versions to be quite similar or rather different. In early mainframe versions there was little opportunity to interact with the system, and commands had to be issued in a batch. Feedback from the system (for example on simple syntax errors) was often slow. In some computer centres there is still a delay while files are processed, and sometimes a long wait in a queue before output is printed.

Because the microcomputer is not usually shared between users at any one time, such delays do not occur. Individual commands can be entered interactively and immediate feedback is provided about certain types of error. An especially useful aspect of SPSS/PC+ is the facility it has for providing on-line information via the **HELP** command. On the other hand, the micro versions have limitations of size which provide certain restrictions (in particular, SPSS/PC+ files can contain no more than 200 variables). Also, certain procedures available in SPSS^X are not available in SPSS/PC+ (including, at the time of writing, **GUTTMAN**, **RELIABILITY**, **NONPAR CORR** and **BOX-JENKINS**).

Certain procedures have different commands when initiated within

SPSS/PC+ and SPSS^X. Thus the following are equivalent:

SPSS/PC+   SPSS^X

CORRELATION	**PEARSON CORR**
JOIN	**MATCH FILES, ADD FILES**
MEANS	**BREAKDOWN**
N	**N OF CASES**

In SPSS/PC+ (but NOT in SPSS^X):

- each command must end with a terminator (usually **.**) or a blank line
- string variables must be written exactly (including spaces)
- all data must be specified in the file before the first procedure
- any command keyword can be truncated to just the first three characters
- files are named directly (not through a file handle)
- logical variables are not supported – a logical **expression** must be entered.

In SPSS^X, **OPTIONS** and **STATISTICS** are commands – in SPSS/PC+ they are subcommands associated with various procedures. Also, certain of the **OPTIONS** available in SPSS^X are not available in SPSS/PC+ (for example, with the procedures **CORRELATION** and **CROSSTABS**), and certain defaults are different between the two versions.

I.2 SPSS/PC+ and SPSS/PC

SPSS/PC+, as the + suggests, is an enhanced version of the earlier SPSS/PC. Although it contains several improvements and additions (including **REVIEW** and **JOIN**), certain programs which were included in the original version (**HILOG**, **FACTOR**, and **CLUSTER**) have now been removed from the base system and are available (together with other programs) in the optional *Advanced Statistics* package.

The changes in using the versions are, for the most part, minor. Full details are provided in Appendix G of the *SPSS/PC+ Manual*. Those who are familiar with SPSS/PC will find that almost all of their normal commands work as well with SPSS/PC+. There are fewer restrictions (for example in the permitted length of labels). Certain little-used **OPTIONS** have been removed, but there is more freedom to control certain aspects of the system, for example with the **SET** command. A command can be terminated in SPSS/PC+ with a terminator (usually **.**) **or** with a blank line (this is achieved simply by pressing <RETURN>. This option was not available in the early version of SPSS/PC.

Appendix J: Brief Guide to SPSS/PC+ Commands, Options and Statistics

If SPSS/PC+ is used infrequently, the user will need to be reminded of even the most basic rules of syntax of the various commands. The following examples of commands will serve to remind you that when using the **CROSSTABS** procedure the syntax is 'x BY y' whereas when using the **CORRELATION** procedure the syntax is 'x WITH y', etc. The table also indicates some of the most frequently used **STATISTICS** and **OPTIONS**.

This guide is not comprehensive and should be used only as a memory aid after you have worked through the relevant chapters of this book.

ANOVA   (Performs analysis of variance)

```
ANOVA VAR1 VAR3 TO VAR5 BY AGEGROUP(1,4) VAR8 (1,3).
```

This will produce four two-way anovas. Each of the dependent variables (*VAR1, VAR3, VAR4* and *VAR5*) will be subjected to separate analysis. In each analysis there are four levels of the independent integer variable *agegroup* and three levels of the second independent integer variable *VAR8*.

OPTIONS: 2. Suppress labelling   9. Regression approach
STATISTICS: 3. Display counts and cell means

COMPUTE   (Creates a new variable, producing values for each case)

```
COMPUTE VAR101=VAR1 + VAR2 + VAR3.
COMPUTE VAR102=VAR1 + (VAR2/VAR3)**2.
```

CORRELATION   (Performs Pearson product-moment correlations)

    CORRELATION VAR1 VAR2 VAR3.

This will produce three correlations (1–2, 1–3, 2–3)

    CORRELATION VAR1 TO VAR10 WITH VAR2 TO VAR10.

This will produce a matrix of 90 correlations, including 9 in which a variable is correlated with itself (the value therefore will be 1.000)

OPTIONS: 2. Pairwise deletion   3. Two-tailed probability
                4. Write matrix results file
                5. Display includes count and probability values
STATISTICS: 1. Display of variable counts, means and
                   standard deviations

CROSSTABS   (Crosstabulates data on two or more integer variables)

    CROSSTABS VAR1 VAR2 VAR3 BY VAR4 VAR8.

This will produce six crosstabulation tables (1–4, 1–8, 2–4, 2–8, 3–4, 3–8). The first-named variables (*VAR1, VAR2, VAR3*) will form the rows of the tables, the second-named variables (*VAR4, VAR8*) the columns.

OPTIONS: 2. All labels suppressed
                18. Display of all cell information
STATISTICS: 1. Chi-squared test

DATA LIST   (Fixed format) (Lists variables and indicates their position within the dataset)
1. For data included in the file (i.e. inline data):

    DATA LIST /VAR1 1-2 VAR2 3-5 VAR3 6 VAR4 7 ... VAR32 80 / VAR33
    1-2.

2. For data to be read from an external data file "gender.dat":

    DATA LIST FILE= "gender.dat" /VAR1 1-2 VAR2 3-5 VAR3 6 VAR4 7.

3. For data to be read from an external matrix file "gender.mat":

    DATA LIST MATRIX FILE= "gender.mat" / VAR1 1-2 VAR2 3-5 VAR3 6.

DISPLAY   (Provides information about variables in the active file)
1. DISPLAY.
2. DISPLAY ALL.
3. DISPLAY VAR8 TO VAR14.
2 and 3 will provide fuller information about variables than will 1.

EXPORT   (Produces a portable ASCII version of the active file)

```
EXPORT OUTFILE= "gender.por".
```

(produces a portable file "gender.por" containing all of the current active file – to produce a file of only part of the contents use **DROP** or **KEEP**)

```
EXPORT OUTFILE= "gender.por"/DROP= VAR15 to VAR19.
```

The **TO** convention **is** permitted on the **DROP** or **KEEP** subcommand.

FREQUENCIES   (Displays frequency counts for values of single variables)

```
FREQUENCIES VAR1 TO VAR26.
```

OPTIONS: No OPTIONS are available with FREQUENCIES, although subcommands include BARCHART and HISTOGRAM.

STATISTICS: For this procedure (only), statistics are requested not by number but by name. The default statistics are: **MEAN STDDEV MINIMUM MAXIMUM**. Others which may be requested include **RANGE MODE MEDIAN VARIANCE**. The user may also request **ALL**. The statistics subcommand therefore takes the following form:

```
/STATISTICS=DEFAULT RANGE VARIANCE.
```

(this would request the default statistics plus the two others).

GET   (Reads a system file)

```
GET FILE= "gender.sys".
```

(A **DROP** subcommand may also be included, and the **TO** convention **is** permitted.)

HELP   (Provides on-screen information about types of file, commands, etc).

HELP ALL.	Provides list of basic HELP topics
HELP CORRELATION.	Provides general help on COR-RELATION
HELP CORRELATION OPTIONS.	Lists CORRELATION OPTIONS

IF   (Transforms or creates a variable if a logical condition is fulfilled)

```
IF (VAR1 GT 3) GROUP=2.
IF (VAR3 NE VAR4) VAR101=2.
IF (VAR3 EQ 1 OR VAR3 EQ 2 OR VAR3 EQ 3) VAR58=VAR3 + 12 - VAR6.
```

In the last of these examples, if *VAR3* for a case equals 1, 2 or 3 then (and only then) a variable *VAR58* for that case is computed by adding the value 12 to the *VAR3* value and then subtracting the value of *VAR6*. If *VAR3* for the case does not equal 1, 2 or 3 then *VAR58* for that case is system missing.

IMPORT   (Reads a portable ASCII file containing data and data dictionary)

```
IMPORT FILE= "gender.por".
```

(**DROP** or **KEEP**, **RENAME** and **MAP** subcommands may also be included.)

The TO convention is permitted on a **DROP** or **KEEP** subcommand.

INCLUDE   (Used to read a command file, including a file which itself contains an **INCLUDE** command – such nesting can be five levels deep)

```
INCLUDE "gender.def".
INCLUDE "test.sps".
```

JOIN (Combines up to five files – one can be the active file, all others must be system files)

With the keyword **MATCH**, a **JOIN** command combines different variables for the same cases:

```
JOIN MATCH FILE= "genpre.sys"
/FILE= "genpost.sys".
```

With the keyword **ADD**, a **JOIN** command combines different cases which have data for the same variables:

```
JOIN ADD FILE= "genmale.sys"
/FILE= "genfem.sys".
```

With either **MATCH** or **ADD**, the subcommands **DROP** or **KEEP**, **RENAME**, **MAP** and **BY** may also be included. If no **BY** is included in a **MATCH** then cases are matched in the order in which they occur in the two files (first with first, second with second, etc.). If **BY** is included it must be placed as the last subcommand (except for **MAP**) and the key variable used to match the cases (in the following example it is *ID*) must be present in all files to be matched. The files must also be in prior order by the key variable (thus it might be necessary to **SORT** them before the **JOIN**). A **BY** subcommand with **ADD** can be used to determine the order in which cases will occur in the new file. The current active file must be indicated by an asterisk:

```
JOIN MATCH FILE=*
/DROP VAR34 VAR38 VAR44
/FILE= "genpost.sys"
/DROP VAR134 VAR138
/BY ID
/MAP.
```

The **TO** convention is **NOT** permitted on a **DROP** or **KEEP** subcommand

---

LIST (Displays the values of all (or specified) variables for cases in the active file)

```
LIST ALL.
LIST VAR1 TO VAR16.
```

Data on specific cases only can also be requested:

```
LIST ID VAR1 TO VAR16
/CASES= FROM 1 to 20.
```

MEANS   (Displays means, standard deviations and group counts for dependent variables defined by independent variable(s))

```
MEANS VAR1 VAR2 VAR4 BY SEX AGEGROUP.
```

VAR1 VAR2 and *VAR4* are the dependent variables; *SEX* and *AGE-GROUP* are the independent variables. Six tables are requested. Compare with:

```
MEANS VAR1 VAR2 VAR4 BY SEX BY AGEGROUP.
```

SEX groupings are made and, within these, *AGEGROUP* subgroups. Thus mean values will be given for old males, young females, etc. This command requests three tables.

OPTIONS: 3. Suppress all labels   12. Display group variances
STATISTICS: 1. One-way analysis of variance

MISSING VALUE   (Declares a value – one only – which indicates that the datum for that variable is missing for that case)

```
MISSING VALUE VAR1 VAR4 TO VAR20 (9) VAR2 VAR3 (99).
```

N   (A case selection command. Limits the number of cases in the working file to the first *N* cases)

```
N 50.
```

NPAR TESTS   (Performs a number of non-parametric tests)

```
NPAR TESTS KENDALL = VAR1 VAR2 VAR4 VAR6.
```

(*See* Chapter 13 for details of each test)

OPTIONS: 2. Listwise deletion   3. Sequential pairing of data
                                        from two related groups
STATISTICS: 1. Count, mean, standard deviation, minimum
                 and maximum

ONEWAY (Performs one-way analysis of variance)

```
ONEWAY VAR1 VAR3 VAR5 BY AGEGROUP(1,3).
```

A number of *post-hoc* multiple contrast tests can be requested by the **RANGES** subcommand:

```
ONEWAY VAR1 BY AGEGROUP(1,3)
/RANGES=DUNCAN.
```

OPTIONS: 2. Listwise deletion   4. Write matrix
                     7. Read matrix

The matrices referred to in OPTIONS 4 and 7 include counts, means and standard deviations.

STATISTICS: 1. Group descriptive statistics

PLOT (Produces two-dimensional graphical plots of bivariate distributions)

```
PLOT PLOT VAR4 WITH VAR7.
```

No **OPTIONS** (or **STATISTICS**) are available with PLOT, but there are a number of subcommands including:

TITLE – provides heading
VERTICAL – provides label for vertical axis
HORIZONTAL – provides label for horizontal axis
FORMAT – specifies the type of plot to be produced

PROCESS IF (Produces a temporary selection of cases to be included in the next procedure)

PROCESS IF selects cases according to the value of each case on a specified variable.

```
PROCESS IF (VAR3 GE 3).
PROCESS IF (VAR3 LT 2).
```

Logical operators (**AND, OR, NOT**) cannot be included in **PROCESS IF** commands.

The following relational operators are permitted:

EQ (or =), **GT** (or >), **LT** (or <), **NE** (or <>), **GE** (or >=), **LE** (or <=)

RECODE   (Revises the coding scheme for an existing variable)

```
RECODE VAR1 (1,2,3=1) (4=2) (5,6,7=3).
```

Permitted keywords include **LO, HI, THRU** and **ELSE**:

```
RECODE SALARY (LO THRU 999=1) (1000 THRU 1999=2) (ELSE=3).
```

The keywords **HI, LO** and **ELSE** should be used with extreme care to avoid the inadvertent recoding of missing values as valid.

REPORT   (Produces lists of cases and summary statistics)

REPORT allows the user to tailor the appearance of summary tables, including titles, column widths and footnotes. There are no **OPTIONS** or **STATISTICS**, but many subcommands are available. *See* Chapter 14. Note also that there are several other ways of producing tables for inclusion within reports. The optional SPSS/PC+ **TABLES** enhancement allows the user to produce high quality camera ready tables. Alternatively the user can edit a listing file containing output from other procedures (e.g. **MEANS**) and then tailor the appearance using an editor program (e.g. a word processor).

REVIEW   (Allows the editing of text files, including data files, data definition files, the log file and the listing file)

REVIEW.	Produces two-window display of current listing (top) and log (bottom) files
REVIEW LOG.	Produces display of current log file
REVIEW "gender.dat".	Opens new or existing file space to create or edit text file

Review can be entered from within SPSS/PC+ or directly from **DOS**.

Pressing <F1> provides on-screen guide to **REVIEW** commands.

SAMPLE   (Draws temporary random selection of cases to be processed in the next procedure)

```
SAMPLE .25.
SAMPLE 100 from 253.
```

SAVE (Saves current active file as a system file)

```
SAVE.
```
Produces system file with default name SPSS.SYS (unless default has been changed)

```
SAVE OUTFILE = "gender.sys"
```
Produces system file named "gender.sys"

System files cannot be edited.

A **DROP** subcommand may also be included, but there is *NO* **KEEP** subcommand. The **TO** convention **is** permitted on the **DROP** subcommand.

SELECT IF (Selects – according to a logical criterion – certain cases to remain in the active file

The selection remains in effect until the end of the session or until another file is made active.

SELECT IF selects cases according to the value of each case on a specified variable.

```
SELECT IF (VAR3 GE 3 OR VAR4 GE 3).
SELECT IF (VAR3 LT 2).
SELECT IF (VAR3 GT (VAR4*10)).
```

Thus relational operators (**EQ, GE, NE,** etc.) and logical operators (**AND, OR, NOT**) can be included in **SELECT IF** commands. The logical expression is evaluated. If true, the case is selected. If false, the case is not selected.

SET (Changes any of a wide range of optional features of the SPSS/PC+ operation) Use SHOW command to display current settings.

```
SET PRINTER ON.
SET MORE OFF.
SET LISTING = "gend.lis".
SET PRINTER ON / LOG= "gend.log" / BEEP OFF.
```

SHOW (Displays current settings for a wide range of optional features of the SPSS/PC+ operation)

```
SHOW.
```

To change any of the features displayed use the **SET** command (see above).

SORT   (Re-orders cases in the active file according to the value of each case on one or more – up to 10 – sort keys)

`SORT BY AGE.`	Sorts cases in ascending order of age
`SORT BY AGE(D).`	Sorts cases in descending order of age
`SORT BY AGE(A) SCORE(A) HT(D)`	Sorts cases first by age, in ascending order. Cases in the same age category are further sorted by score, also in ascending order. Any cases with the same age and score are sorted by descending order of HT.

SUBTITLE   (*See* **TITLE**)

TITLE and **SUBTITLE**   (Changes default heading and adds subtitle to pages of output)

By default, each page of SPSS/PC+ output includes the date, a page number and a centre title 'SPSS/PC+'. This title is changed if the user issues a **TITLE** command followed by up to 58 characters followed by a terminator (.). The use of quotation marks will allow printing of both upper and lower case and the inclusion of apostrophes, etc.

`TITLE "First stage of Michael's Correlational Analysis".`

Similarly, **SUBTITLE** causes a user-supplied label to be included beneath the title line:

`SUBTITLE "Male Subjects' Data".`

Subtitles are not centred but start in the first column. To cancel a title or subtitle enter the single command word only.

T-TEST (Performs an independent *t*-test or paired *t*-test to test the difference between the means of two samples)

```
T-TEST GROUPS= sex(1,2)/ VARIABLES = VAR1 TO VAR12.
```

This command would be used to perform 12 **independent** *t*-tests on the two groups coded 1 and 2 for the variable *sex*.

```
T-TEST PAIRS= VAR1 VAR2 VAR3 VAR4.
```

This command would be used to perform 6 **paired** *t*-tests; the values of all valid cases in the active file would be compared (in pairwise fashion) on each possible pair of variables in the list.

```
T-TEST PAIRS= VAR1 VAR2 WITH VAR3 VAR4.
```

This command would be used to perform 4 **paired** *t*-tests; each of *VAR1* and *VAR2* would be compared with each of *VAR3* and *VAR4*.

OPTIONS: 2. Listwise deletion   3. Variable labels suppressed
5. WITH interpreted to execute sequential pairing

STATISTICS: No additional statistics are available. The default display includes t-value, degrees of freedom, and two-tailed probabilities. Counts, means, standard deviations and standard error are calculated for each variable or group. Other statistics are also provided.

VALUE LABELS (Provides labels for some or all of the values which a variable may take)

```
VALUE LABELS SEX 1 "Female" 2 "Male" /AGE 1 "Under 20" 2 "20+"
/VAR37 TO VAR41 1 "YES" 2 "NOT SURE" 3 "NO"
/VAR23 VAR25 VAR27 1 "Strongly Agree" 7 "Strongly Disagree".
```

VARIABLE LABELS (Provides labels for variables)

```
VARIABLE LABELS VAR1 TO VAR6 "Attitudes to men"
/VAR7 TO VAR12 "Attitudes to women".
```

Index